T0066838

A SEA IN FLAMES

ALSO BY CARL SAFINA

The View from Lazy Point
Nina Delmar: The Great Whale Rescue
Voyage of the Turtle
Eye of the Albatross
Song for a Blue Ocean

A SEA IN FLAMES

The Deepwater Horizon Oil Blowout

CARL SAFINA

BROADWAY PAPERBACKS
NEW YORK

Published in the United States by Broadway Paperbacks,
an imprint of the Crown Publishing Group,
a division of Random House, Inc., New York.
www.crownpublishing.com

Originally published in hardcover in a slightly different form in the United States by
Crown Publishers, an imprint of the Crown Publishing Group,
a division of Random House Inc., New York, in 2011.

Library of Congress Cataloging-in-Publication Data
Safina, Carl, 1955–
A sea in flames : the Deepwater Horizon oil blowout / Carl Safina.
p. cm.
1. BP Deepwater Horizon Explosion and Oil Spill, 2010—Environmental
aspects. 2. BP Deepwater Horizon Explosion and Oil Spill, 2010—Social
aspects. 3. Oil spills—Mexico, Gulf of. I. Title.
GC1221.S24 2011
363.738'20916364—dc22 2010051455

ISBN 978-0-307-88736-8
eISBN 978-0-307-88737-5

Book design by Elizabeth Rendfleisch
Cover design by David Tran
Cover photography by U.S. Coast Guard/Getty Images

First Paperback Edition

146028962

To the memories of the people who died.

To their families.

To those who survived.

To the creatures that suffered.

To those who anguished.

To those who did their best.

And to those who continue asking what will come out of this well.

CONTENTS

Preface: Know Before You Go ix

PART ONE

DISASTER CHAIN

1

Blowout! 3

April 45

Déjà Vu, to Name but a Few 57

PART TWO

A SEASON OF ANGUISH

65

Mayday 67

Late May 93

Early June 119

High June 127

Late June 148

Like a Thousand Julys 196

Late July 222

PART THREE

AFTERMATH

237

Dog Days 239

Late August 259

Early September 269

The New Light of Autumn 274

References 301

Acknowledgments 341

Index 345

Preface

Know Before You Go

Crucial mistakes, disastrous consequences, the weakness of power, unpreparedness and overreaction, the quiet dignity of everyday heroes. The 2010 Gulf of Mexico blowout brought more than oil to the surface.

This is not just a record of a technological event. It's also a chronicle of a season of anguish and panic, deep uncertainties, and the emotional topography of the blowout. It is the record of an event unfolding, a synthesis of personal experience, news, rumors, and the rapidly shifting perspectives about how bad things were—and how bad they were not.

There are roughly three parts to this event, and to this book: what caused this particular well to blow out; the varied technological, biological, and emotional responses during the months the oil was flowing; and a little more calmness, clarity, and insight after the flow of oil was stopped.

I've chosen to convey my impressions as they occurred over a season that was intense, chaotic, and seemingly interminable. In the turmoil, it was easy to form the wrong impressions and follow blind alleys. And I did.

Over the months, information and understanding improved significantly. Later, after the flow of oil was stopped, we calmed down, and those with cooler heads began to see more clearly.

This book is not a definitive treatise; it's a portrait. The story will continue unfurling. Some aspects, we'll never fully understand.

In trying my best to get it right, I am sure that nearly all of what I've written is reasonable, most of it is true, and some of it is wrong. It's not less than that, and not more.

It's easy to criticize people in charge. It's much harder to be the person in charge. I was angry at the Coast Guard for weeks, until I began to realize that its ability to respond was largely dictated by the laws that confined it. If officials such as Admiral Thad Allen rankled me at times, it may say more about me than about them. But it remains part of the portrait of this whole event.

In truth, such people deserve not just our admiration but also a little slack. During the blowout, perfection wasn't an available option. I've left my first impressions in place to show how my perceptions changed as my initial rage—and I felt plenty of rage—subsided. Admiral Allen, as the most visible federal official and the man in charge, gets the brunt of my exasperation. But he never fully deserved it. I could not have done the job he did.

Admiral Allen, Dr. Jane Lubchenco, and others in our government gave us their very best under months of intense pressure, heavy responsibility, and public scrutiny. They were doing a nearly impossible job on behalf of us all. I didn't always appreciate that right away, especially during my summer travels through the Gulf region, when I was often both angry and grief-stricken. In truth, they deserve our thanks and praise.

But it's not all about them. It's about us. We all contributed to this event, and we're all trapped in the same situation. We all use too much gasoline and oil, because we've painted ourselves into a corner when it comes to energy.

For clarity I have lightly cleaned up or slightly condensed some of the verbatim testimony and quotes. Verbal exchanges during the hours leading up to and including the initial disaster on the drilling rig derive from recollections of those who endured that trauma. Because they are subject to the fog of crisis, some testimony conflicts; we may never know how to resolve those contradictory recollections.

In the end, this is a chronicle of a summer of pain—and hope. Hope that the full potential of this catastrophe would not materialize, hope that the harm done would heal faster than feared, and hope that even if we didn't suffer the absolute worst, we'd still learn the big lesson here.

We may have gotten two out of three. That's not good enough. Because: there'll be a next time.

Carl Safina
Stony Brook, New York
November 2010

PART ONE

DISASTER CHAIN

BLOWOUT!

April 20, 2010. Though a bit imprecise, the time, approximately 9:50 P.M., marks the end of knowing much precisely. A floating machinery system roughly the size of a forty-story hotel has for months been drilling into the seafloor in the Gulf of Mexico. Its creators have named the drilling rig the Deepwater Horizon.

Oil giant BP has contracted the Deepwater Horizon's owner, Transocean, and various companies and crews to drill deep into the seafloor forty-odd miles southeast of the Louisiana coast. The target has also been named: they call it the Macondo formation. The gamble is on a volume of crude oil Believed Profitable.

Giving the target a name helps pull it into our realm of understanding. But by doing so we risk failing to understand its nature. It is a hot, highly pressurized layer of petroleum hydrocarbons—oil and methane—pent up and packed away, undisturbed, inside the earth for many millions of years.

The worker crews have struck their target. But the Big Payback will cut both ways. The target is about to strike back.

A churning drill bit sent from a world of light and warmth and living beings. More than three miles under the sea surface, more than two miles under the seafloor. Eternal darkness. Unimaginable pressure. The drill bit has met a gas pocket. That tiny pinprick. That pressure. Mere bubbles, a mild fizz from deep within. A sudden influx of gas into the well. Rushing up the pipe. Gas expanding like crazy. Through the open gates on the seafloor. One more mile to the sea surface.

The beings above are experiencing some difficulty managing it. A variety of people face a series of varied decisions. They don't make all the right ones.

Explosion.

Fireball.

Destroyed: Eleven men. Created: Nine widows. Twenty-one fatherless kids, including one who'll soon be born. Seventeen injured. One hundred and fifteen survive with pieces of the puzzle lodged in their heads. Only the rig rests in peace, one mile down. Only the beginning.

Blowout. Gusher. Wild well. Across the whole region, the natural systems shudder. Months to control it. Years to get over it. Human lives changed by the hundreds of thousands. Effects that ripple across the country, the hemisphere, the world. Imperfect judgment at sea and in offices in Houston, perhaps forgivable. Inadequate safeguards, perhaps unforgivable. No amount of money enough. Beyond Payable.

Deepwater exploration had already come of age when, in 2008, BP leased the mile-deep Macondo prospect No. 252 for $34 million. By 1998 only two dozen exploratory wells had been drilled in water deeper than 5,000 feet in the Gulf of Mexico. A decade later, that number was nearly three hundred.

With a platform bigger than a football field, the Deepwater Horizon was insured for over half a billion dollars. The rig cost $350 million and rose 378 feet from bottom to top. On the rig were 126 workers; 79 were Transocean employees, 6 were BP employees, and 41 were subcontractors to firms like Halliburton and M-I Swaco. None of the Deepwater Horizon's crew had been seriously injured in seven years.

Operations began at Macondo 252 using Transocean's drilling platform Marianas on October 6, 2009. The site was forty-eight miles southeast of the nearest Louisiana shore and due south of Mobile, Alabama. As the lessee, BP did the majority of the design work for the well, but utilized contractors for the drilling operation. Rig owner Transocean was the lead driller. Halliburton—formerly headed by Dick Cheney, before he became vice president of the United States under George W. Bush—was hired for cementing services. Other contractors performed other specialized work.

The initial cost estimate for the well was approximately $100 million. The work cost BP about $1 million per day.

They'd drilled about 4,000 feet down when, on November 8, 2009, Hurricane Ida damaged Marianas so severely that the rig had to be towed to the shipyard. Drilling resumed on February 6, 2010, with BP having switched to Transocean's Deepwater Horizon rig.

By late April, the well would be about $58 million over budget.

Being a deepwater well driller—what's it like? To simplify, imagine pushing a pencil into the soil. Pull out the pencil. Slide a drinking straw into that hole to keep it open. Now, a little more complex: your pencil is tipped not with a lead point but with a drilling bit. You have a set of pencils, each a little narrower than the last, each a little longer. You have a set of drinking straws, each also narrower. You use the fattest pencil first, make the hole, pull it out, then use the next fattest. And so on. This is how you make the hole deeper. At the scale of pencils-as-drills, you're going down about 180 feet, and the work is soon out of sight. As you push and remove the pencils, you slide one straw through another, into the deepening hole. You have a deepening, tapering hole lined with sections of drinking straw, with little spaces between the hole and each straw, and between the sections of straw. You have to seal all those spaces, make it, in effect, one tapering tube, absolutely tight.

And here's why: the last, narrowest straw pokes through the lid of a (very big) pop bottle with lots of soda containing gas under tremendous pressure. As long as the lid stays intact and tight, there's no fizz. But only that long. Everyone around you is desperate for a drink of that pop, as if they're addicted to it, because their lives depend on it. They're in a bit of a hurry. But you have to try to ignore them while you're painstakingly working these pencils and straws. And you'd better keep your finger on the top of the straw, or you're going to have a big mess. And you'd better seal those spaces between sections of straw as you go down, or you're going to have a big mess when you poke through that lid. And before you take your finger off the top of the straw, you'd better be ready to control all that fizz and drink all that pop, because it's coming up that straw. And if, after poking a hole in this lid that's been sealed for millions of years, you decide you want to save the soda for later, then you'd better—you'd better—have a way to stopper that straw before you take your finger off. And you'd better have a way to block that straw if the stopper starts leaking and the whole thing starts to fizz. If it starts to fizz uncontrollably, and you can't regain control, you can get hurt; people can die.

The real details beggar the imagination of what's humanly and technologically possible. Rig floor to seafloor at the well site: 5,000 feet of water, a little under one mile. Seafloor to the bottom of the well: about 13,360 feet—two and a half miles of drilling into the seabed sediments. A total of 18,360 feet from sea surface to well bottom, just under three and a half miles.

Equally amazing as how deep, is how narrow. At the seafloor—atop a well 2.5 miles long—the top casing is only 36 inches across. At the bottom it's just 7 inches. If you figure that the average diameter of the casing is about 18 inches, it's like a pencil-width hole 184 feet deep. Nine drill bits, each progressively smaller, dig the well. The well's vertical height gets lined with protective metal casings that, collectively, telescope down its full length. At intervals, telescoping tube of casing gets slid into the well hole. The upper casing interval is about 300 feet long. Some of the lower ones, less than a foot across, are 2,000 feet long. The uppermost end of each casing will have a fatter mouth, which will "hang" on the bottom of the previous casing. You will make that configuration permanent with your cementing jobs. The casings and drill pipes are stored on racks, awaiting use. Casings are made in lengths ranging from 25 to 45 feet; the drill pipe usually comes in 30-foot joints. They are "stacked" in the pipe racking system. You assemble three at a time and drop approximately 90 feet in, and then repeat. When you get ready to put the casing in, you pull all the drill pipe out. Rig workers also remove the drill pipe from the hole every time the drill bit gets worn and needs changing or when some activity requires an open hole. Pulling the entire drill string from the hole is called "making a trip." Making a trip of 10,000 feet may take as long as ten or twelve hours. When you want to start drilling some more, you have to reassemble the drill pipe and send it down.

On drillers' minds at all times is the need to control the gas pressure and prevent gas from leaking up between the outside of the casings and the rock sides of the well. At each point where the casing diameter changes, the well drillers must push cement between the casings and the bedrock wall of the well. This cements the casings to the well wall. It controls pressure and eliminates space.

Drillers continually circulate a variety of artificial high-density liquid displacements or drilling fluids, called "mud," between the drill rig and the well. The circulating fluid is sent down the drill pipe. It causes the drill bit to rotate, then leaves the drill bit and comes up to the surface, carrying the rock and sand that the drill bit has ground loose. Because the well is miles deep, the fluid creates a miles-high column of heavy liquid. (The drilling fluid is heavier than water. Imagine filling a bucket with water and lifting it; then imagine that the bucket is three miles tall. It's heavy.) That puts enormous downward pressure on the entire well bore. As the drill digs deeper, the drilling "mud" formulation is made heavier to neutralize the higher pressures in the deepening depths. But that heavier fluid can exert so much pressure on the shallower reaches of the well (where the ambient pressure is less) that it can fracture the rock, damage the well, seep away, and be lost into the rock and sand. Steel casings can protect weaker sections of rock and sand from these fluid pressures.

Because things fail and accidents happen, a 50-foot-high stack of valves sits on top of the well on the seafloor. Called a "blowout preventer," it is there to stop the uncontrolled release of oil and gas when things go wrong in a well. If something goes seriously wrong below, the valves pinch closed, containing the pressure. The blowout preventer is relied on as the final fail-safe.

Designs vary. This rig had a 300-ton blowout preventer manufactured by Cameron International. A blowout preventer's several shutoff systems may include "annulars," rubber apertures that can close around any pipe or on themselves; "variable bore rams," which can seal rubber-tipped steel blocks around a drill pipe if gas or oil is coming up outside it; "casing shear rams" or "super shear rams," designed to cut through casing or other equipment; and "blind shear rams," designed to cut through a drill pipe and seal the well. Blind shear rams are the well-control mechanism of last resort. Though often designed with redundant equipment and controls, blowout preventers can fail. On occasion, they have. Neither casing shear rams nor blind shear rams are designed to cut through thick-walled joint connections between sections of drill pipe. Such joints may take up as much as 10 percent of a pipe's length. So having redundant shear rams ensures that there is always one shear ram that is not aligned with a tool joint.

The drilling fluid is the primary stopper for the whole well. If you're going to remove that stopper, you'd better have something else to hold the pressure. Usually, that something else is several hundred feet of cement. On the night of the explosion, as rig workers were preparing to seal the well for later use, drillers were told to remove the drilling fluid and replace it with plain seawater—in essence, to pull out the stopper. The cement did not hold. And in the critical moment, the blowout preventer failed. The consequent gas blast was the blowout.

That's what went wrong. But so many things had gone wrong before the blowout that assistant well driller Steve Curtis had nicknamed it "the well from hell." Curtis, thirty-nine, a married father of two from Georgetown, Louisiana, was never found.

❧

Right from the start—beginning with Hurricane Ida forcing the Marianas rig off the well location—various things didn't proceed as planned, or struck people as risky.

The Deepwater Horizon, built at a cost of $350 million, was new in February 2001. In September 2009, it had drilled the deepest oil well in history—over 35,000 feet deep—in the Gulf of Mexico's Tiber Field.

It was a world-class rig, but it was almost ten years old. The wonderful high-tech gadgets that were state of the art in 2001 did not always function as well in 2010. Equipment was getting dated. Old parts didn't always work with new innovations. Manufacturers changed product lines. Sometimes they had to find a different company to make a part from scratch.

The world has changed a lot since the rig was built. So has software. More 3-D, a lot more graphics. Drillers sit in a small room and use computer screens to watch key indicators. Depth of the bit, pressure on the pipe, flows in, flows out. But on this job, the software repeatedly hit glitches. Computers froze. Data didn't update. Sometimes workers got what they called the "blue screen of death." In March and April 2010, audits by maritime risk managers Lloyd's Register Group identified more than two dozen components and systems on the rig in "bad" or "poor" condition, and found some workers dismayed about safety practices and fearing reprisals if they reported mistakes.

Risk is part of life. And it's part of drilling. Yet drilling culture has changed, with much greater emphasis on safety than in the past. Many people still working, however, came up the ranks in a risk-prone, cowboy "oil patch" culture. A friend of mine who worked the Gulf of Mexico oil field in the 1970s says, "It was clear to me that I was way underqualified for what I was doing. Safety didn't get you promoted. They wanted speed. If we filled a supply boat with five thousand gallons of diesel fuel in twenty-five minutes, they'd rather you disconnect in a big hurry and spill fifty gallons across the deck than take an extra three minutes to do it safe and clean. I'd actually get yelled at for stuff like that. Another thing that was clear: if you could simply read or write, you could pretty much run the show. They actually gave oral exams to workers who couldn't read. I was still a kid, but pretty soon I was put in charge of a supply boat because I could read and write. That was the culture then." Another friend, now a tug captain, says, "Never in the four years I worked the rig did I hear anyone say, 'Let's wait for better sea conditions.' We were always dragged into situations we didn't want to be in, doing things I didn't think were safe. Now it's a lot better. It used to be the Wild West out there."

When you pump drilling fluid down the well, it comes out the bottom of the drill pipe and circulates up between the drill pipe and the wall of the well, and comes back to you. For every barrel of drilling fluid you push down, you'd better get a barrel back. If you get more—that's really bad, because gas and oil are coming up in your fluid. If you get less—that's really bad, too. Drillers call it "lost returns." It means the returning fluid has lost some of its volume because fluid is leaking into the rock and sand of the well's walls, sometimes badly. Sometimes there are fractures in the rock and the fluid's going there. When it's leaking like that, you can't maintain the right pressure in the well to tamp down the pressure of oil and gas that wants to come up from below.

In a March 2010 incident, the rig lost all of its drilling fluid, over 3,000 barrels, through leaks into the surrounding rock and sand formation into which they were drilling.

BP's onshore supervisor for this project, John Guide, later testified, "We got to a depth of 18,260 feet, and all of a sudden we just lost complete returns."

BP's senior design engineer, Mark Hafle, was questioned on this point:

Q: "Now, lost returns, what does that mean in plain everyday English?"

Hafle: "While drilling that hole section we lost over 3,000 barrels of mud."

Three thousand barrels is a lot of barrels. At over $250 per barrel for synthetic oil-based mud, that's $750,000.

A high-risk pregnancy is one running a higher than normal risk for complications. A woman with a high-risk pregnancy needs closer monitoring, more visits with her primary health-care provider, and more careful tests to monitor the situation. If BP can be called the birth parent, this well was a high-risk pregnancy.

Several times, the well slapped back with hazardous gas belches called "kicks," another indication that the deep pool of hydrocarbons did not appreciate being roused from its long sleep.

At around 12,000 feet, the drill bit got stuck in rock. The crew was forced to cut the pipe, abandon the high-tech bit, and perform a time-consuming and costly sidetrack procedure around it to continue with the well. The delays cost a week and led to a budget add-on of $27 million.

The work had fallen forty-three days behind schedule, at roughly $1 million a day in costs. At a "safety meeting," the crew was informed that they'd lost about $25 million in hardware and drilling fluid. Not really safety information. More pressure to hurry.

High-risk pregnancy, added complications. On April 9, 2010, BP had finished drilling the last section of the well. The final section of the well bore extended to a depth of 18,360 feet below sea level, which was 1,192 feet below the casing that had previously been inserted into the well.

At this point, BP had to implement an important well-design decision: how to secure the final 1,200 or so feet and, for eventual extraction of the petroleum, what kind of "production casing" workers would run inside the protective casing already in the well. One option involved hanging a steel tube called a "liner" from the bottom of the previous casing already in the well. The other option involved running one long string of steel casing from the seafloor all the way down to the bottom of the well. The single long string design would save both time (about three days) and money.

BP chose the long string. A BP document called the long string the "best economic case." And though officials insist that money was not a factor in their decisions, doing it differently would have cost $7 to $10 million more.

BP's David Sims later testified, "Cost is a factor in a lot of decisions but it is never put before safety. It's not a deciding factor."

Sims was John Guide's supervisor. Guide described the long string design as "a win-win situation," adding that "it happened to be a good economic decision as well."

Guide insisted that none of these decisions were done for money.

Q: "With every decision, didn't BP reduce the cost of the project?"

Guide: "All the decisions were based on long-term well-bore integrity."

Q: "I asked you about the cost of the project. Didn't each of these decisions reduce the cost, to BP, of this project?"

Guide: "Cost was not a factor."

Q: "I didn't ask if it was a factor. I asked if it reduced the cost. It's a fact question, sir. Did it not reduce the cost, in each case?"

Guide: "All I was concerned about was long-term well-bore integrity."

Q: "I just want to know if doing all these decisions saved this company money."

Guide: "No, it did not."

Q: "All right; what didn't save you money?"

Silence.

Q: "Which of these decisions that you made drove up the cost of the project, as opposed to saving BP money? Can you think of any?"

Guide: "I've already answered the question."

Q: "What was the answer?"

Guide: "These decisions were not based on saving BP money. They were based on long-term well-bore integrity."

Some people called the long string design the riskier of two options. Greg McCormack, director of the University of Texas at Austin's Petroleum Extension Service, calls it "without a doubt a riskier way to go."

But others disagree. Each of the two possible well casing designs represented certain risk trade-offs. One called for cement around casing sections at various well depths, providing barriers to any oil flowing up in the space between the rock and casing. The other called for casing sections seamlessly connected from top to bottom with no outside barriers except the considerable bottom cement. Investigators would later focus lasers on this aspect of the well design for weeks after the well blew. The cost savings led many to believe that this was a cut corner that resulted in the blowout. Months later, however, it became clear that this decision was not a direct cause of the disaster.

Final hours. In the eternal darkness of the deep sea, the well is dug, finished. All that's needed: just seal the well and disconnect. The plan was for a different rig to come at some later date and pump the oil for sale.

At BP's onshore Houston office, John Guide is BP's overall project manager for this well. He has been with BP for ten years, has overseen more than two dozen wells. Mark Hafle is BP's senior design engineer. With twenty-three years at BP, Hafle created much of the design for this well. Brian Morel, a BP design engineer involved in many of the key meetings and procedures in the final days, splits his time between Houston and the rig. Out on the drilling rig itself, BP's supervisors, titled "well site leaders," are often called the "company men." They oversee the contractors. Because a drilling rig operates twenty-four hours a day, BP has two well site leaders aboard, working twelve-hour shifts: Don Vidrine and Bob Kaluza. Vidrine is in his sixties. Kaluza in his fifties. Vidrine has been with the rig for a while. Kaluza is new.

Because the Deepwater Horizon was both a drilling rig and a vessel, rig owner Transocean has two separate leadership roles. When moving, the rig is under the authority of the captain; when stationary at the well site, an offshore installation manager, or OIM, is in charge. Jimmy Harrell, OIM, managed the drilling. He'd been with Transocean since 1979 and on the Deepwater Horizon since 2003. Curt Kuchta was the Deepwater Horizon's captain.

Managers play an important part in the decision process, but the drilling team executes the plan. At the top of the drilling personnel chart are the "tool pusher," who oversees all parts of the drilling process, and the driller, who sits in a high-tech, glass-paneled control room called the "driller's shack" and leads the actual work. Many people work under the direction of the tool pusher and driller.

On duty on the evening of April 20 were Transocean's tool pusher Jason Anderson and driller Dewey Revette. Thirty-five years old, Jason had worked on the Horizon since it launched, in 2001, and was highly respected by his crewmates. At home before the explosion, Jason had been concerned about putting his affairs in order. He wrote a will and gave his wife, Shelley, instructions about things to do if anything were to "happen to him." Jason told his father that BP was pushing the rig operators to speed up the drilling. In telephone calls from the rig before the explosion, Jason told Shelley he could not talk about his concerns because the "walls were too thin," but that he would tell her about them later, when he got home. Jason had just been promoted to senior tool pusher. He had been due to leave for his new post aboard the Discoverer Spirit on April 14, but was persuaded to stay aboard the Deepwater Horizon for one more week. He was scheduled to be helicoptered to his new job at 7:00 A.M. on April 21. By then, he had died in the explosions and the rig was an inferno.

At the closing of a well, it might seem you'd want the team members most familiar with the well and one another to be present. But approaching the critical juncture of closing up the well they'd been drilling for months, one of BP's company men, with thirty-three years of experience as a well site leader, was sent off the rig to take his mandatory biannual well-control certification class. Just four days before the explosion, his replacement, Bob Kaluza, appeared on the rig. A *Wall Street Journal* article said of Kaluza, "His experience was largely in land drilling," and he told investigators he was on the rig to "learn about deep water," according to Coast Guard notes of an interview with him. We don't have a better feel for Kaluza because he has exercised his Fifth Amendment right not to provide testimony that could incriminate himself.

Transocean's onshore manager responsible for the Horizon, Paul Johnson, testified that he was troubled by the timing in BP's switch of well site leaders. "I raised my concerns," he noted. "I challenged BP on the decision. We didn't know who this gentleman was. I wasn't making any assumptions on him, I just—I heard he come from a platform, so I was curious about his deepwater experience in a critical phase of the well. They informed me that Mr. Kaluza was a very experienced, very competent well site leader, and it wouldn't be a concern."

Kaluza showed a tendency toward appropriate caution, but the simple fact that he was new seemed to get in the way. At a critical juncture, Kaluza was uncertain enough about a crucial procedure called a "negative pressure test" that he sought out Leo Lindner, a drilling fluid specialist with the company M-I Swaco. Lindner, who'd worked on the rig for over four years, was in charge of the different types of fluids used during the negative test, an important role.

"Mr. Bob Kaluza called me to his office," Lindner testified. "He wanted to go over the method. I briefly explained to him how the rig had been conducting their negative tests and he just wanted—." Lindner interrupted himself to note, "Bob wasn't the regular company man on the Horizon."

So, competence aside, there were working dynamics, team cohesion. It was a time for familiar faces and an almost literally well-oiled team. But it felt like BP was taking out its quarterback during the fourth quarter of a playoff game.

And on the morning of April 20—the day the rig exploded—BP

engineer Brian Morel departed the rig, creating space for visiting company VIPs. Wrote one industry analyst later, "Let's face it; the timing of that VIP visit was terrible. It could not have been at a worse time."

A difficult pregnancy, new doctors, altered procedures: BP decided to turn this exploratory well into a production well. Usually, the purpose of an exploratory well is to learn about the geological formation and what the oil and gas–bearing production zone contains. Then the well is closed out. Engineers use the information to decide where to drill a production well, perhaps in a nearby spot. If you decide to turn an exploratory well into a production well, you obviously save a fair amount of drilling expense. But is an exploration crew going to be familiar with production technology? BP's drilling and completion operations manager David Sims testified that the decision was not a major technical issue. Yet veteran well site leader Ronnie Sepulvado who'd been on the rig for eight and a half years had a different take: "We're in the exploration group, so we hardly ever set production strings. We did maybe a handful of wells that was kept for production."

Added complications. The oil and gas—in the pay zone, or "production zone"—lay between 18,051 and 18,223 feet. The well was drilled to below the zone, to 18,360 feet below the sea surface, which allowed cement to be placed under the oil and gas reservoir as well as around it.

Because this was an exploratory well, the idea was to find the oil, then seal the well shut so a different rig could later tap it for commercial production. Cement is the main barrier for preventing the pressurized oil and gas from entering the well. So it was crucial that the cement job at the bottom of the well absolutely seal off the oil and gas reservoir from the well casing. A bad cement job could let oil and gas into the well.

The environment at that depth means cementing is not a matter of getting a few bags of concrete from the hardware store. Temperatures and pressures at the bottom of a well like this—it's hotter than boiling, 240° Fahrenheit—make cementing a highly technical endeavor, requiring calculations and tests to select several chemical mixtures, which will be used in layers.

Earlier heavy losses of drilling fluid told technicians that they could be into very loose rock and sand. If you are nervous about a soft zone, you also worry that when you insert cement to seal the well bottom, your cement may ooze into the loose stuff. This complicated the cementing deliberations. John Guide: "The biggest risk associated with this cement job was losing circulation. That was the number one risk."

If you're worried that the well walls may be so porous that they'll suck in cement pumped under pressure, you might add some nitrogen gas to the cement mixture, to get it to form foamy bubbles; this would prevent the cement from leaking into the loose spots.

From the well's training resources document: "Foamed cement is more expensive than regular cement and it works better than regular cement in some applications. One of the advantages is that the bubbles stiffen the wet cement so that it is less prone to being lost into a zone or being invaded by fluids in a zone. A remote analogy is that when a sink is drained after washing dishes, the water flows out the drain while the soap bubbles remain in the sink." But, the document notes, while foamed cement is good at sealing off shallow areas, "use of nitrogen foam is less common for deep high-temperature, high-pressure zones."

Halliburton cement specialist Jesse Gagliano first proposed including nitrified cement. After some back-and-forth, BP agreed. But because nitrified cement is usually used for shallower jobs, the depth created concern on the rig.

Transocean offshore installation manager Jimmy Harrell: "That nitrogen, it could be a bad thing. If it gets in the riser, it will unload the riser on you. . . . Anything can go wrong."

There were three parts to the cement and three formulations. "Cap cement" topped the cement in the space between the casing and the oil-bearing rock and sand formation of the well's sides. Below that, the nitrified "foamed cement" filled the rest of the narrow space outside the casing and along the formation. "Tail cement" filled the "shoe track" at the bottom and was used inside the lower part of the casing itself.

So in various ways this was going to be a difficult cement job. As late as the afternoon of April 14, BP was still reconsidering the chosen long string casing design, with its heavier reliance on the integrity of the cement deep at the well bottom. And the porous surrounding rock was on everyone's mind. Cement has to be pumped in under some degree of excess pressure in order to fully fill the gap and get a good bond to the rock and sand on one side and to the outside of the production casing on the other. You need enough pressure both to keep the hydrocarbons contained and to force the cement against the sides, but too much pressure will inject the cement into the sand, and you'll lose it. The team spent days determining how to approach the cement job. BP engineer Mark Hafle testified to this: "We were concerned that the pore pressure and frac gradient was going to be a narrow window to execute that cement job. That's why we spent five days." BP's Brian Morel apologized to a colleague for asking yet another question about the design in an April 14 e-mail that he ended with this resonant comment: "This has been a nightmare well." Hafle added, "This has been a crazy well for sure."

When BP won the lease to this piece of seabed, it held an in-house contest to name it. The winner, "Macondo," came from the mythical town hewn from a "paradise of dampness and silence" in Gabriel García Márquez's novel *One Hundred Years of Solitude.* In the novel, Macondo is an accursed place, a metaphor for the fate awaiting those too arrogant to heed its warning signs. What had seemed a nice literary allusion now carries ominous portent.

More complications. Part of Jesse Gagliano's task was to model the cement's likely performance in this well and design a procedure that would get the cement to the proper locations. On April 15, he discovered some problems. This space between the casing and the wall of this well was very narrow. And the previous experience with lost drilling fluid indicated soft walls, requiring a low cement-pumping rate.

These conditions contributed to a model predicting that if the casing moved too close to one side of the well-bore wall, drilling fluid could get left behind, creating pockets or channels where the cement would not distribute uniformly. That is, it wouldn't fill in all of the space it needed to fill.

To prevent a casing from getting too close to one side of a well bore, drillers slip flexible metal spring devices called "centralizers" over the casing so that it will stay centered in the well bore. By keeping the casing centered, centralizers help achieve good, even, thorough cementing between the casing and the well's geological wall. In this case, BP had six centralizers. That number concerned Gagliano. On April 15 Gagliano e-mailed BP saying he'd run different scenarios "to see if adding more centralizers will help us."

BP's Brian Morel replied, "We have 6 centralizers. . . . It's too late to get any more to the rig. Our only option is to rearrange placement of these centralizers. . . . Hopefully the pipe stays centralized due to gravity."

But Jesse Gagliano continued his calculations. He determined that twenty-one centralizers should create an acceptably safe cement flow.

And it wasn't really too late. On April 16, BP engineering team leader Gregg Walz e-mailed BP project manager John Guide, saying that he'd located fifteen more centralizers that could be flown to the rig in the morning with "no incremental cost" for transporting them. "There are differing opinions on the model accuracy," he wrote to Guide, "but we need to honor the modeling." He added, "I apologize if I have overstepped my bounds."

The centralizers made the helicopter trip to the rig.

But Guide expressed dismay at these particular centralizers' design, the addition of new pieces "as a last minute decision," and the fact that it would take ten hours to install them. He wrote, "I do not like this," adding that he was "very concerned about using them."

Walz backed off.

Later that afternoon BP's Brian Morel wrote to his colleague Brett Cocales, "I don't understand Jesse's centralizer requirements."

Cocales replied, "Even if the hole is perfectly straight, a straight piece of pipe in tension will not seek the perfect center of the hole unless it has something to centralize it." And then he added this: "But who cares, it's done, end of story, will probably be fine and we'll get a good cement job."

That was on April 16. It seems to suggest a certain willingness to add risk.

That's not how BP's managers saw it. Guide later testified: "It was a bigger risk to run the wrong centralizers than it was to believe in the model."

But months later in September, BP's own internal investigation concluded, "The BP Macondo team erroneously believed that they had received the wrong centralizers."

In late July 2010, examiners from BP contractors Anadarko, Transocean, and Halliburton questioned Guide on his decisions.

Q: "That left you several days to get whatever centralizers you felt might be needed."

Guide: "I didn't feel they were needed."

Q: "So what you're telling me is that there was just no discussion among you between you and Mr. Walz about just waiting for the right centralizers? None, zip, zero, true?"

Guide: "That subject never came up."

Q: "You still had time between the 16th and the 20th—"

Guide: "Well, we didn't know if we could find them. That subject never came up."

Q. "Sir, can you tell us the number of times, that you have personal knowledge of, that BP did not follow the recommendations of Halliburton in connection with the cementing of any of its jobs, if any?

Guide: "I don't know of any."

Well, perhaps we know of one. Halliburton's Gagliano accepted BP's decision and, on April 17 and 18, developed the specific procedure for pumping the cement. Gagliano created and sent one final cementing model out to the team on the evening of April 18.

The model would later cause a firestorm for a particular page that no one at BP seems to have looked at. That page said that using only six centralizers would likely cause channeling; it also noted: "Based on analysis of the above outlined well conditions, this well is considered to have a SEVERE gas flow problem." But with twenty-one centralizers, it added, "this well is considered to have a MINOR gas flow problem."

This report was attached to an e-mail sent to Guide on April 18, but it went unopened because the casing with just the six centralizers was already down the hole. Although BP had had days to get the centralizers, it was now too late to read the e-mail predicting severe gas-flow problems. Guide later testified: "I never knew it was part of the report."

The cement job will fail. But a few months later, in September 2010, BP's own investigation will conclude, "Although the decision not to use twenty-one centralizers increased the possibility of channeling above the main hydrocarbon zones, the decision likely did not contribute to the cement's failure."

That's BP's executives exonerating themselves, so season it with a grain of salt. But numerous industry analysts think centralizers are not the smoking gun. We'll get back to that question later, but for now, it's important to understand the distances involved. The recommended twenty-one centralizers were meant to keep the bottom 900 feet of casing evenly centered in the well. If the workers had had all twenty-one, they would have put fifteen above the span containing the oil and gas, four in the zone that held the oil and gas, and two below that zone.

BP placed the six centralizers so as to straddle and bisect the 175 vertical feet of oil and gas–bearing sands deep in the well, at depths of around 18,000 feet. They placed two centralizers above the oil and gas zone, two in the zone, and two below it.

But even if centralizers won't be the smoking gun, the e-mail exchanges over the centralizers convey the sense that the BP team isn't treating this endeavor with the utmost care. When red flags go up, BP's decision makers seem rushed, rather than thorough.

BP e-mails suggest that its personnel believed that any problem with cement could be remediated with additional cement. And actually, that's often what's done; well cement jobs sometimes do fail. The reason why they fail is seldom precisely ascertained. Usually the failure is not catastrophic and the fix is to pump more cement in, then test it again. For this reason, the industry has developed several ways of testing the soundness of cementing jobs.

But detecting problems assumes, of course, that the cementing job will be properly tested.

The crew did their cementing over a five-hour period, starting at 7:30 P.M. on April 19. When they finished, at around 12:30 A.M., the calendar had turned over to April 20.

Testing. More complications: when wells lose drilling fluid—as this one did weeks earlier—one possible solution is to send down a special mixture of fluids to block the problem zones in the well bore. Think of the stuff made to spray into a flat tire to seal it enough to get you home. A batch of this mixture is called a "kill pill." It is a thick, heavy compound (16 pounds per gallon, compared to 14.5 for drilling fluid and 8.6 for seawater).

Two weeks before the accident, when the rig had its serious 3,000-barrel loss of drilling fluid, the fluid specialists made up a kill pill and pumped it down to the problem zone. It didn't seem to work, so they mixed up another batch: 424 barrels of a combination of two materials. But just as they were preparing to send this second kill pill down the hole, the losses stopped.

They now had a thick, unused 424-barrel kill pill sitting in an extra tank, taking up space on the rig. To dispose of it they had two options: take it onto shore and treat it as hazardous waste or use it in the drilling process. The second choice would allow them to skirt the land-based disposal process and dump the compound directly into the ocean.

The drilling fluid specialists got the bright idea of using the unused kill-pill material in a "spacer." A spacer is a distinct fluid placed in between two other fluids. When you're pushing different fluids down a well, you'll often decide to use a spacer between the different fluids—between displacement fluid and drilling fluid, for instance—so that they won't mix and so you can keep track of where things are. A spacer also creates a marker in the drilling flow, which allows the rig team to watch the fluid returns, to ensure that flow in equals flow out.

Because BP didn't want to have to dispose of the thick kill-pill material, they mixed it with some other fluid to create a spacer. BP's vice president for safety and operations, Mark Bly, later said that using such a mixture was "not an uncommon thing to do." The rig's drilling fluid specialist, Leo Lindner, put it differently, saying, "It's not something that we've ever done before." At a government hearing in August, BP manager David Sims was asked if he had ever used a similar mixture as a spacer. "No, I have not," Sims said.

Down the hatch it goes. Just like that.

Q: "What if you hadn't used it that way, what would the rig have had to do; hazardous waste disposal, right?"

Lindner: "Yes."

Q: "When these pills are mixed, have you ever heard anybody characterize it as looking like snot?"

Lindner: "It wasn't quite snotty."

Q: "But it was close?"

Lindner: "It was thick. It was thick, but it was still fluid."

Q: "So it was very viscous?"

Lindner: "Yes."

Q: "And really the only reason for putting those two pills down there was just to get rid of them; is that your understanding?"

Lindner: "To my knowledge—well, it filled a function that we needed a spacer."

Chief engineer Steve Bertone later recalled that after the explosion, "I looked down at the deck because it was very slick and I saw a substance that had a consistency of snot. I can remember thinking to myself, 'Why is all this snot on the deck?' "

Back on the rig, Transocean installation manager Jimmy Harrell outlines the well-closing procedure. BP's company man (later testimony is conflicting as to whether Kaluza or Vidrine was speaking) suddenly perks up. Interrupts. Says, "Well, my process is different. And I think we're gonna do it this way." Chief mechanic Douglas Brown will later testify that BP's company man said, "This is how it's going to be," leading to a verbal "skirmish" with Transocean's Jimmy Harrell, who left the meeting grumbling, "I guess that's what we have those pincers for" (referring to the blowout preventer). Harrell will later testify that he was alluding to his concerns about risks inherent in the cementing procedure, but would say, "I didn't have no doubts about it." He'll claim he had no argument but that "there's a big difference between an argument and a disagreement." Chief electronics technician Mike Williams, seated beside BP's company man, will later recall, "So there was sort of a chest-bumping kind of deal. The communication seemed to break down as to who was ultimately in charge."

This is certainly not the time for chest bumping or blurred authority. If you're gonna release the parking brake, you'd better agree on who's gonna be in the driver's seat. And whoever grabs the wheel better know how to drive.

High-risk pregnancy enters labor. To determine if the cement job has worked and the well is sealed, rig operators can choose from several tests. On the Deepwater Horizon the engineers decide to do two kinds of pressure tests. In a "positive pressure test," they introduce pressure in the well; if it holds, it means nothing's leaking *out* from the well into the rock. They do this test on the morning of April 20, between about 11:00 a.m. and noon, roughly eleven hours after the cement job ends. It goes well; it seems nothing's leaking out.

But the reason nothing's leaking out may be that there's pressure from oil and gas pushing to get in. So the engineers prepare to do a "negative pressure test." A negative test is a way of seeing if pressure is building in the well, indicating that gas and oil are leaking in. That could mean the cement has failed.

To do a negative test, they close the wellhead, then reduce the downward pressure on the well by replacing some heavy drilling fluid with lighter water. Then they look at pressure gauges. If the pressure increases, hydrocarbons are entering, exerting upward pressure from below. What they want to see is zero pressure.

Until the negative pressure test is performed successfully, the rig crew won't remove the balance of the heavy drilling mud that stoppers the well; that's their foot on the brake.

Between 3:00 P.M. and 5:00 P.M., about fifteen hours after the cement job was finished, they start reducing the pressure by inserting seawater into the miles-long circulating-fluid lines. To make sure the drilling fluid and the seawater don't mix, they precede the seawater with a spacer. The spacer they use contains that extra kill-pill material, the "snot." And though a typical amount of spacer is under 200 barrels, this time it's over 400 barrels because, remember, they're trying to get rid of that leftover stuff.

There are various places all this fluid is getting to, because there are various lines and pipes going into and out of the blowout preventer. One such line is called the "kill line." Another is the drill pipe.

A little before 5:00 P.M., they work for a while to relieve any residual pressure and are looking for the fluid to stabilize at zero pressure, indicated by a reading of zero pounds per square inch, or psi. They've got the pressure down to 645 psi in the kill line, but it's at 1,350 psi in the drill pipe. So they try bleeding the system down, venting off some of that pressure. They achieve zero in the kill line. The drill pipe retains 273 psi. They need zero.

Over six minutes right around 5:00 P.M., the drill pipe pressure increases from 273 psi to 1,250.

The engineers tighten the blowout preventer's rubber gasket and add 50 barrels of heavier-than-water fluid.

So the lines are filled with a variety of different fluids snaking through in segments: there's a stretch of drilling fluid, or "mud," a stretch of the unusual spacer material, followed by plain seawater. At this point in the circulation of the various fluids, the spacer—the "snot"—should be above the blowout preventer. But some of it has found its way into the blowout preventer and has entered one of the lines being tested.

The engineers see pressure building in the drill pipe, zero pressure in the kill line. They're unsure what to make of that, so they repeat the test procedure several times. From shortly after 5:00 to almost 5:30, they get the pressure in the drill pipe down a little, from 1,250 to almost 1,200.

The Deepwater Joint Investigation panel asked Dr. John R. Smith, whose PhD is in petroleum engineering, to describe a negative test:

"If it's a successful test, there's no more fluid coming back. You've got a closed container. There's no hole in the boat. There's no fluid leaking in through the wall of the container or the casing. It just sits there. If you have an unsuccessful test, external pressure is leaking through the wall of the system somewhere. Through the wall of the casing, past the casing hanger seals, up through the float equipment in the casing—. Somewhere there's a leak from external pressure into the system. You'd expect to continue to see some fluid coming back."

To reach for an analogy: if you did a negative test in a swimming pool, you'd empty the pool. If the pool stayed dry, you'd have a successful test. If you had a problem, the pool would start filling itself through leaks in its walls. A well 18,360 feet deep is a bit trickier to test than a swimming pool. Even though you're reducing the pressure, you still have to keep enough downward pressure on the well to control any oil and gas that might start entering. And you must check specific pipes for indications of pressure.

Wells come in many different sizes and shapes and pipe setups. So, somewhat surprisingly, there isn't a "standard" negative test.

Q: Do you know if there's any standard negative test procedure that the industry follows?

Dr. John Smith: I was unable to find a standard.

The Minerals Management Service's permit specified that this negative test be conducted by monitoring the kill line above the blowout preventer. John Guide: "And that was really the only discussion, was to make sure that we did it on the kill line so that we would be in compliance with the permit."

Drilling fluid specialist Leo Lindner had spent four years on Deepwater Horizon.

Q: "And what is a good negative test?"

Lindner: "Where you don't have any pressure up the kill. Of course . . . I haven't been a witness to that many negative tests."

And that's another thing: negative tests are not routine on exploratory wells. The Deepwater Horizon mainly drilled exploratory wells. This well was unusual, because it was an exploratory well that was being converted to a *future* production well that would later be reopened and tapped. Not all the crew were familiar with all these steps and procedures.

Dr. John Smith: "Before they ever started the test, they've got enormously high pressure on the drill pipe." That should have been, he noted, "a warning sign right off the bat."

Leo Lindner: "They decided to go ahead and try to do the first negative test. They bled off some pressure from the drill pipe and got fluid back. They attempted it again and got fluid back."

But as Dr. Smith had said: "If it's a successful test, there's no more fluid coming back."

The crew had been replacing heavier fluid with seawater. But Lindner was sufficiently worried by the initial results that at around 5:00 P.M., he ordered his coworker to stop pumping drilling fluid off the rig. He wanted to keep his foot on the emergency brake.

At 5:30, Transocean's subsea supervisor Chris Pleasant comes on duty in the drill shack. "My supervisor was explaining to me that they had just finished a negative test. Wyman Wheeler, which is the tool pusher, was convinced that something wasn't right. Wyman worked to 6:00 P.M. By that time his relief come up, which is Jason Anderson, which is a tool pusher as well."

It was a bad time to change guards.

But now it's approximately ten minutes till 6:00. Bob Kaluza, the BP company man, tells Jason Anderson, "We're at an all stop."

Kaluza's relief, BP company man Don Vidrine, is scheduled to come on at 6:00 P.M.

Chris Pleasant: "Jason Anderson, he's convinced that it U-tubed. Where that U-tube's at, I don't know. But, you know, I guess we never really had a clear understanding. Anyway Jason is telling Bob that, 'We want to do this negative test the way Ronnie Sepulvado does it.' And Bob tells Jason, 'No, we're going to do it the way Don wants to do it. So, probably five minutes after 6:00 or something Don comes to the rig floor. Him and Bob talks back and forth for approximately a good hour."

They discuss possible causes for the fact that they're reading pressure on the drill pipe but not on the kill line. Don Vidrine believes that if the pressure in the drill pipe was evidence of a surge of gas deep in the well, they would be seeing similar pressure in the kill line.

Later question: "Based on industry standard ways of reading negative tests, you're looking for something pretty simple, right? A zero on the drill pipe and a zero on the kill line; right?"

Dr. John Smith: "Right."

Q: "And if you don't see that, you need to be very concerned; right?"

Smith: "Yes." Dr. Smith further says, "We know there's all this heavy spacer mud stuff in the well below the blowout preventer. Likely that mixture is what's going back up into the kill line, holding the pressure back."

Smith adds, "We're doing a test with a line that's got this dense stuff in it. So, the symptoms are a successful test, but the reality is—it's not a test at all. My opinion."

In other words: the only reason they've got zero pressure showing on the line they're relying on is that the thick spacer material has gotten in; the line is clogged.

After thirty minutes of staring at zero pressure on the kill line, the team is convinced that they've completed a successful negative test. Never mind that the drill pipe has 1,400 psi on it. They've convinced themselves that this was due to something Jason Anderson was calling a "bladder effect."

BP well site leader trainee Lee Lambert was later examined on this point.

Q: "What was Mr. Anderson saying about the bladder effect? Can you tell us?"

Lambert: "That the mud in the riser would push on the annular and transmit pressure downhole, which would in turn be seen on your drill pipe."

Q: "Was Mr. Anderson explaining why they were seeing differential pressure on the drill pipe versus the kill line?"

Lambert: "Yes."

Q: "Okay. And did anyone say anything or disagree with Mr. Anderson's explanation?"

Lambert: "I don't recall anybody disagreeing or agreeing with his explanation. At the time it did make sense to me. My lack of experience—. After learning things after the incident, it did not make sense to me, because the kill line and the drill pipe are open up to the same annulus, so in theory should see the same pressure."

Q: "And since then have you had an opportunity to study this so-called 'bladder effect'?"

Lambert: "I have not found any studies on the bladder effect."

In September 2010, BP's internal investigation concluded: "According to witness accounts, the toolpusher proposed that the pressure on the drill pipe was caused by a phenomenon referred to as 'annular compression' or 'bladder effect.' The toolpusher and driller stated that they had previously observed this phenomenon. After discussing this concept, the rig crew and the well site leaders accepted the explanation. The investigative team could find no evidence that this pressure effect exists."

After the negative pressure test, Vidrine tells Bob Kaluza, "Go call the office. Tell them we're going to displace the well." They're about to remove their fluid and replace it with seawater. Poised on a mountaintop, over an oil volcano, they're about to release the brake.

They're in a bit of a hurry. But what about the cement job; had it cured correctly already? The negative test helped convince them that it had. But that was only because the kill line was clogged and they chose to explain away the pressure they were seeing on the drill pipe.

The industry standard for judging the success of cement work, to best try to ascertain whether the cement is bonding to everything properly, is called a "cement bond log test." Halliburton, which did the cement job, will later tell a Senate committee that a cement bond log test is "the only test that can really determine the actual effectiveness of the bond between the cement sheets, the formation and the casing itself."

Of course, because Halliburton did the cement job, its people would like to blame BP for not using the definitive test. They don't want anyone focusing on their cement itself.

Using sonic tools, a cement bond log test makes 360-degree representations of the well and can show where the cement isn't adhering fully to the casing and where there may be paths for gas or oil to get in. In reality, even a cement bond log test is not perfect. But it is the best test going.

Perhaps the most skilled people to do a cement bond log test work for the rig-servicing company Schlumberger. They're on the rig on the morning of April 20, ready to get to work.

BP decides instead to just rely on the pressure tests and other indicators that say that all's well with the well. BP tells the Schlumberger workers that their services won't be needed after all, and arranges for them to leave.

John Guide explains: "Everyone involved on the rig site was completely satisfied with the job. You had full returns running the casing, full returns cementing the casing. Saw lift pressure, bumped the plug, floats were holding. So really all the indicators you could possibly get. So it was outlined ahead of time in the decision tree that we would not run a bond long if we saw these indicators. So the decision was made to send the Schlumberger people home."

As the Schlumberger folks board a helicopter and lift off the rig, oil and gas are already trying to get into the well, pushing hard on the cement. At 11:00 A.M., as the helicopter flies out of sight, eleven men on the rig have eleven hours left to live.

The main critical error was in not recognizing that the drill pipe pressure they were reading during the pressure test indicated that gas was already getting in—and, therefore, that the cement job had failed.

Why did it fail? People will speculate for months. Some will suggest that the cement was not allowed to set adequately before BP began altering the well pressure during the positive and negative pressure tests. Others will see that as irrelevant. Even the time required for the cement to harden at the pressure and temperature deep in the well will be subject to controversy.

Not until September and October did some of the most important pieces of this puzzle start to fit into a clearer glimpse of what happened.

First, as promised, let's revisit the centralizers. In late September 2010, when the relief well finally intersects the original well, it will find no oil outside the casing above the oil-bearing zone in the rock. This will confirm that the oil and gas flowed first out of the sand, then down more than 80 feet outside the casing, then into the well casing and up through 189 feet of "shoe track" cement within the casing. This entire 270-foot run—down outside, then up inside the casing—was supposed to be filled with cement. It's astonishing that the cement in the casing failed. The inner cement was designed to be a solid seven-inch-diameter, 189-foot-long plug.

Though Halliburton had recommended twenty-one centralizers to help ensure a good cement job, BP used only six. But the other fifteen would have been placed *above* the zone bearing the oil and gas. Above that hydrocarbon zone where the other fifteen centralizers would have gone, the engineers poured 791 feet of cement into the gap between the casing and the well wall. That upper cement, above the oil and gas, remained sound. Cement failed in and *below* the main hydrocarbon zone. The part of the cement job that failed was where the centralizers were, and in the reach below them, and *inside* the casing. The flow of gas and oil was not up outside the casing but out of the sand, then *down* to the bottom of the well, then up inside the casing—despite the cement there—and then out to the surface.

This seems to acquit the centralizers. So was there something wrong with the cement formulation itself?

In its September 8, 2010, investigative report, BP will blame the cement. They'll offer several reasons why the cement could have failed: contamination with drilling or spacer fluids, contamination among the three cement parts, or "nitrogen breakout," in which the nitrogen breaks out of the foam and forms big gas pockets.

BP will have a third party make cement samples designed to resemble Halliburton's and test them in a lab. (They could not get cement samples from Halliburton because those are impounded as legal evidence.)

BP's report will claim that Halliburton's foamed cement would have broken down in the well. BP will say Halliburton used a mixture containing 55 to 60 percent nitrogen, and that lab results indicate that "it was not possible to generate a stable nitrified foam cement slurry with greater than 50% nitrogen by volume at the 1,000 psi injection pressure." The investigation team will conclude that "the nitrified foam cement slurry used in the Macondo well probably experienced nitrogen breakout, nitrogen migration and incorrect cement density. This would explain the failure. . . . Nitrogen breakout and migration would have also contaminated the shoe track cement and may have caused the shoe track cement barrier to fail."

Of course, BP's investigation is suspect of pro-BP bias. But even if the BP report is self-serving and finger-pointing, this we do know: the cement did fail.

Just before Halloween 2010, the president's Oil Spill Commission will make its own explosive announcement: Halliburton officials knew weeks before the fatal explosion that its cement formulation had failed multiple tests—but they used the cement anyway.

On March 8, 2010, Halliburton e-mailed results of one failed test to BP, but sent only the numbers; there's no evidence that Halliburton specified that the numbers indicated failed testing. BP had overlooked

Transocean's warning of "severe gas flow," and might also have not understood it had been given information predicting that the cement would fail.

Halliburton altered the testing parameters, but the cement failed several tests. Finally, just days before the blowout, one last test indicated that the cement would remain stable. But Halliburton may not even have received the results of that final test before pouring the cement on April 19. BP definitely did not get notified that one of the formulations tested successfully. In other words, when Halliburton pumped cement into the BP well, both companies apparently possessed lab results indicating that what they were doing was unsafe.

After the blowout, Halliburton will refuse to give its exact cement formulation to BP for independent testing for its September report. But in coming weeks the president's Oil Spill Commission's chief counsel, Fred H. Bartlit Jr., will persuade Halliburton to hand over its cement formulation by reminding its officials that anything they withhold from federal investigators will enlarge the Justice Department's billowing civil and criminal charges. He'll then ask Chevron to create a batch and test the mixture under various conditions. In all nine tests, the cement formulation will prove unstable. The unstable cement solidifies into a firm indictment of Halliburton and BP liability.

At the stupendous pressures acting upon the oil and gas in the surrounding strata, even small cracks in the cement are enough to allow the flow rates that would send 60,000 barrels of oil a day out of the well.

Something called a "float collar" enters the discussion. These flapping one-way valves were situated far down the well casing. Rig workers had a hard time getting them to close. They may not have sealed properly, and the fact that the oil and gas shot up the well casing indicates that the float collar's valves also failed.

Having mistakenly concluded that no hydrocarbons are coming into the well, the workers declare success at 7:55 P.M. on April 20.

At 8:02 they begin displacing all the remaining heavy fluids with seawater. This will take over an hour. They know they're near completion when the spacer comes back up to the surface.

Returning fluids are usually directed into "pits" on the rig. This time, when the spacer reaches the surface, the crew directs the material directly overboard. This was their way, remember, of avoiding the requirement to bring it ashore and dispose of it as hazardous waste. While the material is being dumped overboard, certain flow meters, or "mudloggers," are bypassed. In fact, they may have also been bypassed for much of the day as valuable returning drilling fluids were directed onto a waiting ship, before the spacer was just dumped directly overboard.

BP's September investigation will conclude: "The investigation team did not find evidence that the pits were configured to allow monitoring while displacing the well to seawater. Furthermore, the investigation team did not find evidence that either the Transocean rig crew or the Sperry-Sun mudloggers monitored the pits from 13:28 hours (when the offloading to the supply vessel began) to 21:10 hours (when returns were routed overboard)."

Consultant Dr. John Smith will later testify that bypassing the flow-out meter amounted to "eliminating all conventional well control monitoring methods. That's essentially in direct violation of the Minerals Management Service rules."

There's another major wrinkle. Typical spacers are 180 to 200 barrels in volume, an amount that can be pumped out in fifteen to twenty minutes. But because the crew was trying to get rid of all the unused kill-pill material and bypass the solid-waste requirements by using unwanted material, this spacer was over 400 barrels. That meant that the rig crew had to spend an extra fifteen to twenty minutes or so pumping it overboard.

This extra fifteen minutes occurred between 9:15 and 9:30. Had any crewmates been monitoring the flow through the meters, they would have seen some very irregular pressure and flow readings. Those fifteen minutes, fifteen crucial minutes of not monitoring the volume of their fluids, ended at 9:30 P.M., when they so clearly should have realized they had a problem. Those fifteen minutes could have saved the rig.

Halliburton's cement. M-I Swaco's spacer. Transocean and BP's misinterpretation of the negative pressure test. BP's push to replace all the heavy fluid with seawater. An observation comes to me via this e-mail from a friend: "My ex-brother-in-law was up for the weekend. He was a mud engineer on rigs all over the world, offshore and on. He says there are no excuses, the company man's supposed to be in charge of everything. One thing he was very insistent on is that there's no such noun as 'drill' on the rig. You can have drill bit, drill string, drill pipe, but a drill is what you use to find the lifeboats."

Now the crew is bypassing their monitors as their excess spacer is being dumped overboard.

They'd slowed the pumps at approximately 8:50 P.M. in anticipation of the returning spacer. Slowing meant they should have seen reduced flow coming out, but the flow out actually increased. This was another indication that pressurized oil and gas were entering the well.

Starting at approximately 9:01 P.M., without a change in pump rate, the drill pipe pressure increased from 1,250 psi to 1,350 psi. Another indicator. The pressure should have decreased at this time, not increased, because they were replacing fluid weighing 14.17 pounds per gallon with 8.6-pounds-per-gallon seawater. This increase should have gotten the rig crew's attention.

Over the ten-minute period from 8:58 to 9:08, they gained 39 barrels of fluid, the result of upward pressure in the well.

BP's September report will note: "No apparent well control actions were taken until hydrocarbons were in the riser." In other words, gas had already gotten past the blowout preventer and was rushing the final mile to the surface.

Now the rig crew begins sending returning fluids into a mud-gas separator with limited capacity. They may have thought this was just a "kick," a belch.

In fact, an enormous volume of methane was streaming into the well, shooting upward from miles below, expanding as the surrounding pressure lessened, pushing out the fluid above it, gathering itself into an accelerating blowout. In an awful irony, *this* would have been the time to send all the returning material overboard.

BP's report says that they'd gained 1,000 barrels of liquid volume before anyone tried to activate the blowout preventer. The report adds, "Actions taken prior to the explosion suggest the rig crew was not sufficiently prepared to manage an escalating well control situation."

The gas quickly overwhelms the separator's capacity and mud begins flowing onto the rig floor.

The blowout preventer is attached to the wellhead at the seafloor and to the riser pipe that connects to the rig at the surface. Its many components are often used during routine operations like pressure testing, sealing around drill pipe, and pressure control in the well. It doesn't just sit there unless there's an emergency. And it is tested routinely. It was tested just a few days before the explosion.

The blowout preventer is controlled from the rig through two cables connected to two redundant control "pods"—blue and yellow—and with a hydraulic line. Remotely operated vehicles can also control the blowout preventer by directly operating the pods.

The BP investigation team will conclude that, if the blowout preventer had been closed at any time prior to 9:38 P.M., the flow of hydrocarbons to the riser and up to the surface would have been reduced or eliminated.

They missed it by four minutes.

Remember, by pumping the extra 200 barrels of spacer fluid just to save disposal costs, they'd wasted fifteen minutes.

At 9:42, the crew did try to activate the blowout preventer, tightening a gasket against the drill pipe. At first it did not seal. It does appear that the blowout preventer was sealed around the drill pipe at approximately 9:47. The part of the blowout preventer that they closed simply tightened a rubber gasket against the pipe, a chokehold. What they needed was to close the blind shear rams and sever the pipe—to chop its head off and seal the well.

It was too little. And then, it was too late.

Randy Ezell, a Transocean senior tool pusher, was asleep when his room phone rang. He recalled,

> Well, I hit my little alarm clock light and, according to that alarm clock, it was ten minutes till 10:00. And the person at the other end of the line there was the assistant driller, Steve Curtis. Steve opened up by saying, "We have a situation." He said, "The well is blown out." He said, "We have mud going to the crown." And I said, "Well—." I was just horrified. I said, "Do y'all have it shut in?" He said, "Jason is shutting it in now." And he said, "Randy, we need your help." And I'll never forget that.
>
> And I said, "Steve, I'll be— I'll be right there."
>
> So I put my coveralls on; they were hanging on the hook. I put my socks on. My boots and my hard that were right across that hall in the tool pusher's office. So I opened my door and I remember a couple of people standing in the hallway, but I kind of had tunnel vision. I looked straight ahead and I don't even remember who those people were.
>
> I made it to the doorway of the tool pusher's office when a tremendous explosion occurred. It blew me probably twenty feet against a bulkhead, against the wall in that office. And I remember then that the lights went out, power went out. I could hear everything deathly calm. My next recollection was that I had a lot of debris on top of me. I tried two different times to get up, but whatever it was it was a substantial weight. The third time something like adrenaline had kicked in and I told myself, "Either you get up or you're going to lay here and die." My right leg was hung on something; I don't know what. But I pulled it as hard as I could and it came free. I attempted to stand up. That was the wrong thing to do 'cause I immediately stuck my head into smoke. And with the training that we've all had on the rig

> I knew to stay low. So I dropped back down. I got on my hands and knees and for a few moments I was totally disoriented on which way the doorway was. And I remember just sitting there and just trying to think, "Which way is it?"

Now there was mud shooting out the top of the rig and the loud and continuous *whoosh* of surging gas. More explosions followed, igniting a high-intensity hydrocarbon fire fueled by incoming gas and oil.

Up until the explosions, the crew had two different ways to activate the shear rams that could cut the pipe and seal the well. They did not activate them. Then the explosions damaged the control cables and hydraulic line to the blowout preventer, costing the rig crew the ability to control the blowout preventer.

The loss of connections should have triggered the blowout preventer's automatic emergency mode—and closed the blind shear rams. However, the yellow pod had a defective solenoid and the blue pod's batteries were weak, so the blind shear rams did not activate.

Sensors for fire, gas, and toxic fumes were working; any irregularities appeared on a screen. But their audio alarms were inhibited. This is understandable, but many rigs don't allow it. The Deepwater Horizon had hundreds of individual fire and gas alarms. Having the general alarm go off for local minor problems would cost workers sleep, a safety concern. And people would start ignoring alarms—also a safety concern. The idea was: have a person monitoring the computer, and let them control the general audio alarm. Sound it only when conditions require.

But chief electrician Mike Williams has asserted that inhibiting alarms also prevents the computer from activating emergency shutdown of air vents and power. Such a shutdown could have prevented the rig's diesel generator engines from inhaling the gas and surging wildly.

The over-revving engines send surges of electricity that make lights and computer monitors begin exploding. The engines spark, igniting the gas, triggering explosions.

Transocean's chief mechanic, Doug Brown, knew there was a manual engine shutdown system. He also understood that he was not authorized to activate it. He later said, "If I would have shut down those engines, it could have stopped as an ignition source."

Mike Williams hears loud hissing. Hears the engines revving. Sees his light bulbs getting "brighter and brighter and brighter," knows "something bad is getting ready to happen," hears "this awful whoosh."

He reaches for a door that's three inches thick, steel, fire-rated, supported by six stainless steel hinges. An explosion blows the door from those hinges, throwing him across the shop. When he comes around, he's up against a wall with the door on top of him. He thinks, "This is it. I'm gonna die right here."

When he crawls across the floor to the next door, it too explodes, taking him thirty-five feet backward, smacking him up against another wall. He gets angry at the *doors;* he feels "mad that these fire doors that are supposed to protect me are hurting me." He crawls through an opening. He thinks, "I've accomplished what I set out to accomplish. I made it outside. I may die out here, but I can breathe."

Williams can't see. Something's pouring into his eyes. "I didn't know if it was blood. I didn't know if it was brains. I didn't know if it was flesh. I just knew I was in trouble."

There's a gash in his forehead. He's on one of the rig's lifeboat decks. He's got two functioning lifeboats, right there. But he thinks, "I can't board them. I have responsibilities."

He hears alarms, radio chatter, "Mayday! Mayday!" Calls of lost power. Calls of fire. Calls of man overboard. People jumping from the rig.

Transocean's subsea supervisor Chris Pleasant wants the rig's master, Captain Curt Kuchta, to activate the emergency disconnect system, or EDS. With the blowout preventer unresponsive, the last-ditch response is: disconnect the rig from the pipe that is delivering the gas that's feeding the fire. Kuchta replies, "Calm down! We're not EDSing." Jimmy Harrell, Transocean's man in charge of all drilling operations, has just had his quarters destroyed in explosions while he was in the shower; he now comes running, partially clothed, partially blinded by fine insulation debris. He tells Chris Pleasant to activate the emergency disconnect system. Pleasant tries it. All attempts to disconnect the pipe fail.

Having survived the explosions and freed himself from entrapment in debris, Randy Ezell is trying to get his bearings from where he sits, stunned. "Then I felt something and it felt like air," he later recalled. He says to himself, "Well, that's got to be the hallway. So, that's the direction I need to go. That leads out." Crawling over debris, he makes it to the doorway. But then he realizes, "What I thought was air was actually methane and I could feel, like, droplets; it was moist on the side of my face. So he continues crawling down the dark hallway. Suddenly he puts his hand on a body. He hears a groan. In the dark, Ezell can't see who it is. (It's Wyman Wheeler.) Next, he sees a wavering beam of light. Someone is coming down the hallway, their light going up and down as they duck debris hanging from the ceiling and make their way around jutting walls and over a buckled floor. As the approaching person rounds the corner, Ezell recognizes Stan Carden. While they are pulling debris off of Wyman Wheeler, another flashlight arrives, wielded by Chad Murray. Ezell and Carden ask Murray to find a stretcher while they continue removing debris from Wheeler. Thinking it might be quicker to try to help Wheeler walk out, Ezell helps him to his feet, but, after just a couple of steps with his arm around Ezell's shoulder Wheeler, overcome by pain, says, "Set me down. Set me down." So Ezell lets him back down. Wheeler says, "Y'all go on. Save yourself." To which Ezell replies, "No, we're not going to leave you."

Suddenly Ezell hears another voice saying, "God help me. Somebody please help me." He looks. Where their maintenance office had been, all he sees is a pile of wreckage over a pair of feet. Removing that debris requires the efforts of all three: Ezell, Carden, and Murray. When they get the debris off, they realize it's Buddy Trahan, one of Transocean's visiting dignitaries. Trahan's injuries are worse than Wheeler's, so he gets the first stretcher.

Stan Carden and Chad Murray convey Trahan all the way to the lifeboat station. Ezell stays back. "I stayed right there with Wyman Wheeler because I told him I wasn't going to leave him, and I didn't," he recalled later. "And it seemed like an eternity, but it was only a couple of minutes before they came back with the second stretcher."

Carrying Wheeler outside of the living quarters, Ezell notices that the main lifeboats are gone. Then he notices a few people starting to deploy a raft. The men carrying Wheeler continue down the walkway to the raft and set the stretcher down. "And after several minutes," Ezell will recall, "we had everything deployed and got in the life raft. But the main thing is, Wyman was there, you know—he didn't get left behind."

But unbeknownst to those in the boats, others are left behind.

Mike Williams, who minutes earlier could have had both now-departed lifeboats to himself, watches eight other survivors drop an inflatable raft from a crane.

In weekly lifeboat drills they'd practiced accounting for everyone. There is no longer such a thing as "everyone."

Now left watching are Williams, another man, and twenty-three-year-old Andrea Fleytas. Williams experiences several more blasts that he'll later describe as "Take-your-breath-away explosions. Shake-your-body-to-the-core explosions. Take-your-vision-away explosions."

Fire spreads from the derrick to the deck itself.

Williams sees in Andrea's eyes that she seems resigned to death. He says, "It's okay to be scared. I'm scared, too." She says, "What are we gonna do?" Williams outlines the choice: Burn up or jump down.

From where they are, it's ten stories to a black ocean. Bloodied, backlit by raging fire, Williams takes three steps and jumps feet-first. "And I fell for what seemed like forever," he later recalled. He thinks of his wife, their little girl. "A lotta things go through your mind."

Love conquers all. But only sometimes.

He crashes into the sea and the momentum takes him way, way beneath the surface. He pops up thinking, "Okay, I've made it." But he feels like he's burning all over. He's thinking, "Am I on fire?" He just doesn't know.

He realizes he's floating in oil and grease and diesel fuel. The smell and the feel of it. He sees that the oil that has become the sea's surface beneath the rig is already on fire.

Says to himself, "What have you done? You were dry, and you weren't covered in oil up there; now you've jumped and you've landed in oil. The fire's gonna come across the water, and you're gonna burn up." He thinks, "Swim harder!" Stroke, kick, stroke, kick, stroke, kick, stroke, kick. As hard as he can until he realizes: he feels no more pain. He thinks, "Well, I must have burned up, 'cause I don't feel anything, I don't hear anything, I don't smell anything. I must be dead."

He hears a faint voice calling, and next thing, a hand grabs his lifejacket and flips him over into a boat. Then the boat finds one more survivor. Andrea.

A ship that had been tending the rig, the *Bankston,* retrieves those in lifeboats. Not aboard the *Bankston:* Jason Anderson, of Midfield, Texas, thirty-five, father of two, tool pusher, the supervisor on the floor at the time of the accident who'd worried aloud to his wife and dad about safety on the rig and who'd spoken of the "bladder effect" causing the pressure discrepancies they were seeing; Aaron Dale Burkeen, thirty-seven, of Philadelphia, Mississippi, father of a fourteen-year-old daughter, Aryn, and a six-year-old son, Timothy; Donald Clark, forty-nine, an oil industry veteran married to Sheila, living in Newellton, Louisiana; Steven Ray Curtis, thirty-nine, the driller who'd named this the "well from hell"; Roy Wyatt Kemp, twenty-seven years old, who lived in Jonesville, Louisiana, with his wife, Courtney; Karl Kleppinger Jr., thirty-eight, of Natchez, Mississippi, Army veteran of Operation Desert Storm, leaving behind a wife and son; Keith Blair Manuel, fifty-six, father of three daughters, avid supporter of Louisiana State University sports teams, and engaged to be married to his longtime love, Melinda; Dewey Revette, forty-eight, of State Line, Mississippi, having been with Transocean for twenty-nine years, and leaving a wife and two daughters; Shane Roshto, just twenty-two, of Liberty, Mississippi, husband to Natalie and already father to three-year-old son Blaine Michael; Adam Weise, twenty-four, of Yorktown, Texas, a former high school football star who loved the outdoors; Gordon Jones, twenty-eight, of Baton Rouge. A few days after Gordon died, his widow, Michelle, gave birth to their second son.

Mike Williams says, "All the things that they told us could never happen, happened."

For two days, a fireball. So hot it appears to be melting some of the rig. Which finally sinks.

Accusations: BP's own report will later say, "A complex and interlinked series of mechanical failures, human judgments, engineering design, operational implementation and team interfaces came together to allow the initiation and escalation of the accident. Multiple companies, work teams and circumstances were involved." BP's recap: The cements failed to prevent the oil and gas from entering the well. Staff of both Transocean and BP incorrectly interpreted the negative pressure test by tragically explaining away the pressure they were seeing on one gauge. This led them to release the downward fluid pressure on the well by replacing the heavier fluid with seawater in a well that they falsely believed—because the kill line was clogged with the "snotty" spacer—was not exerting upward pressure. It was. The pressure in the drill pipe, which they chose to ignore, was telling them that the cement had failed. They didn't notice other warning signs because they bypassed gauges and routed displacement fluid and their irregularly concocted spacer overboard. But as gas reached the rig, when the crew might have prevented disaster, they routed the flow to a mud-gas separator whose capacity was soon overwhelmed. Gas flowing directly onto the rig got sucked into generators, causing them to surge and spark, igniting a series of explosions. Fire and gas emergency systems that should have prevented those explosions failed. The blowout preventer should have automatically sealed the well but it, too, failed.

Unlike a tanker running aground and spilling oil—a simple cause-and-effect accident—this is a chain disaster. Each of the distinct failures of equipment and judgment, combined, was required to cause the event. And if any single component had not failed, or had been handled differently, this blowout never would have happened. And we're not done yet, because a failure of preparedness to deal with a deepwater blowout will cost many pounds of cure over the coming months.

APRIL

The Coast Guard's Gulf region chief, Rear Admiral Mary Landry, who, we are told, is "leading the government's response," says, "You're getting ahead of yourself a little when you try to speculate and say this is catastrophic. It's premature to say this is catastrophic."

Though eleven men died and the rig sank.

She's just much too cool. Right off the bat, her statements set my confidence in the Coast Guard on a wobble. I realize immediately that it's going to be one of those events where, from all sides, the truth gets pinballed back and forth among bumpers of spin and flippers of distortion. And other than herself, who's Landry kidding? The immediate impressions: (1) she's still wait-and-seeing; and (2) because of No. 1, the official response is slow.

President Barack Obama says the federal response to the disaster is "being treated as the number one priority." He may think so, but it doesn't feel that way. It seems as though no one is prepared for oil shooting from a mile-deep pipe.

Oddly, just one day after the rig sinks, one major press agency is calling the *Exxon Valdez* spill "vastly bigger than the current one in the U.S. Gulf." That may be because right after the rig sank, Rear Admiral Landry said that no oil appeared to be leaking from the wellhead and nor was there, at the surface, "any sign of a major spill." Landry said at the outset that most of the oil was burning off with the fireball, leaving only a moderate rainbow sheen on the water.

The impression given: fuel oil had *spilled* from the rig, but from the well no oil was *leaking.* She said, "Both the industry and the Coast Guard have technical experts actively at work. So there's a whole technical team here to ensure we keep the conditions stable."

Stable.

Even if this isn't "Obama's Katrina," she sounds like Obama's Michael Brown. Heckuva job, Brownie. And now the real Michael Brown comes out of his hole. President George W. Bush's infamous FEMA chief claims that Obama is purposely dragging his feet, wanting the oil leak to worsen so he can shut down offshore drilling. "This is exactly what they want, because now he can pander to the environmentalists and say, 'I'm gonna shut it down because it's too dangerous,'" Brown says, adding, "This president has never supported Big Oil, he's never supported offshore drilling, and now he has an excuse to shut it back down."

How very odd of Brownie to say that, considering that less than a month ago Obama alienated environmentalists with a blindsiding announcement that he was opening millions of acres for new offshore oil development.

Rewind: March 30, 2010. President Obama lays out an offshore drilling plan. He decides to open 167 million acres of ocean to oil and gas exploration. Candidate Obama had attacked John McCain's proposal to expand offshore drilling, saying, "It would have long-term consequences for our coastlines but no short-term benefits since it would take at least ten years to get any oil. . . . When I'm president, I intend to keep in place the moratorium." With this March announcement, President Obama erases that promise. He ends a longstanding moratorium on oil exploration along the East Coast from the northern tip of Delaware to the central coast of Florida. For environmentalists, this leaves a very bad impression. Something like betrayal. Buried in the announcement—because the administration wants it to look like he's into offshore oil—is that President Obama canceled the plans, scheduled by President G. W. Bush, for four lease sales off Alaska in the Beaufort and Chukchi Seas. But he left 130 million acres of those seas open to exploration. Obama's decision keeps the entire West Coast closed to oil and gas leasing. He also takes Alaska's Bristol Bay off the table from any future speculation, thereby keeping its teeming salmon flowing. Obama: "There will be those who strongly disagree with this decision."

Right again. Greenpeace: "Is this President Obama's clean energy plan or Sarah Palin's?" The envirogroup Center for Biological Diversity says, "All too typical of what we have seen so far from President Obama—promises of change and ultimately, adoption of flawed and outdated Bush policies." Republican congressman John Boehner slams the president anyway for not going far enough; he wants to give Big Oil more access. Piling on from the opposite side, Democratic senator Frank Lautenberg rails, "Giving Big Oil more access to our nation's waters is really a 'kill, baby, kill' policy: it threatens to kill jobs, kill marine life, and kill coastal economies that generate billions of dollars. Offshore drilling isn't the solution to our energy problems. I will fight this policy and continue to push for twenty-first-century clean energy solutions."

Obama says his policy is "part of a broader strategy that will move us from an economy that runs on fossil fuels and foreign oil to one that relies more on homegrown fuels and clean energy." And that's exactly what's needed. But other pieces of the strategy are not in place; Congress isn't into clean energy. So Obama's nuance is lost on everyone. The public impression will cost Obama's administration the moral high ground, and will seem to translate—another impression, admittedly—into weeks of lost momentum and lagging leadership after the rig explosion.

Who likes Obama's announcement? Why, it's Obama's former opponent for the presidency, Senator John McCain, who Twitters, "Drill baby drill! Good move."

And it doesn't exactly sound like pandering to enviros when White House spokesman Robert Gibbs says that sometimes accidents happen; loss of the Deepwater Horizon is no reason to back off the president's recent decision to expand offshore drilling.

Yet that face-saving will turn butt-biting.

And this from rig owner Transocean itself: "The U.S. Coast Guard has plans in place to mitigate any environmental impact from this situation." The world's largest offshore drilling contractor, with a fleet of 140 mobile offshore drilling units, seemingly sees scant need to further trouble itself. And well owner BP says, "We are working closely with BP Exploration & Production, Inc. and the U.S. Coast Guard to determine the impact from the sinking of the rig and the plans going forward."

Between that line and this one, you can read the following: they have no response plan; they're unprepared and have no real idea of what to do.

The Coast Guard searches for the lost men. They'll keep searching "as long as there is a reasonable probability of finding them alive." There was never a reasonable probability of additional survivors, and the horror of eleven killed begins to sink in.

Rear Admiral Mary Landry's statement that there's "no apparent leak" notwithstanding, leaking it is. First estimate: 1,000 barrels per day (a barrel is 42 gallons, so 42,000 gallons). The Macondo well's oil is not leaking from the wellhead, or from the blowout preventer. It's leaking from the former umbilicus between rig and well, a kinked pipe broken off when the rig sank, now lying mangled on the seafloor one mile down. The only shard of luck—if that term applies here—is that the hulking mass of the drilling rig itself did not land atop the wellhead, blocking all access to it.

A remotely operated vehicle is sent to shut the leak at the blowout preventer.

Fails.

A few days ago, the gushing oil was termed "manageable." Now BP executives say at least two to four weeks to get it under control. We don't hear them say "at least." We still can't believe the phrase "four weeks." (That estimate, too, will get several resets. Four months from now, we'll wish it had been "four weeks.") Ideas, anyone?

I did not want to come to the Gulf. It seemed like a good situation to avoid. But I also don't feel that the information I'm getting from officials and the news media is fully reliable. So I want to see things for myself. I've resisted initial predictions of ecological disaster, believing that the natural systems will more or less continue functioning. Yet as you'd guess from a statement like that, I was also in a bit of denial; I could not fathom that stopping the leak might require more than a few days. That just couldn't happen. But the new reality is settling in: this won't be quick. Now a lot will depend on what will, indeed, become a months-long question: How much oil will come out of that hemorrhaging stab wound in the seafloor?

Dawn on the coast. Shell Beach, in southeast Louisiana, is the end of the road, the edge of the marsh, the beginning of the Gulf. A man is sitting in his car next to a stone memorial engraved with the names of dozens of local Katrina victims. I feel like I'm interrupting a private moment just glancing at him, but he says hello and seems to want to talk. He says he's not in the seafood business, but he lives here among the fishing families. "The small-timers in the seafood business," he says, "people make fun of them because they don't know the answers to intelligent questions. Maybe they don't know the name of the First Lady of the United States, but that's not what they care about. What they care about is that their motor starts in the morning and that they go out, and go to work. And out there, they're professionals that nobody can compete with; they're scientists at their jobs."

The oil now raises the question of whether an enduring way of life is truly endangered, or whether an endangered way of life will endure.

The sheriff has already erected a guard post to close the road into Shell Beach because the National Guard is building a dock for ferrying materials. The guy in the booth tells me no cars, but he lets me walk through.

He's tall, thin, black, talkative, fortyish, worried. "This'll make Katrina look like a bad day," he says. "Because Katrina did what it did.

Then you picked up and got back to your way of life. But this, I mean—it's disheartening to say it, but I think the parish has pretty much had it. Once that oil comes in, it won't leave. There'll be no people workin' for a *long* time. Years. And every time it rains, we'll get a sheen. And every time we get a sheen, they'll shut fishin'. These fishermen makin' money, spendin' money. That's gone. There's *nothing* that's gonna replace that income. I mean, you probably had ten, twenty thousand sacks of oysters goin' outta this place a day. There's one crab buyer I know, he probably averages thirty, forty thousand pounds a day when the crabs are running good. That's just one buyer. See what I'm sayin'? The average working fisherman that fishes hard, I'd say they gross at least a hundred thousand dollars a year. Minus expenses, but BP can't replace their income forever. Not the rest of their lives."

He tells me to look around at the places that sell bait, fishing gear. "These boats. See what I'm sayin'? That's all gone. You look around at these weekend fishing houses. This is serious money. The average place back here, they got five hundred thousand dollars invested. There's people had camps on order; they already canceled. What are you gonna build a fish camp for, if you can't catch fish? But some's got hundreds of thousands already invested in a new place they won't be able to give away."

He says the parish still runs a tax deficit of $1 million a year. Aftereffects of Katrina. "A lot of people had to leave the parish. Some people say, 'Oh, try to get a Walmart—.' You can have all the stores you want. People got the same amount of money to spend no matter how many stores you have. We can't draw from anywhere anymore. The whole surrounding area was devastated. Before, a lot of the people came from the Ninth Ward to shop. We just don't have that anymore; it's depressed all through there. This oil's gonna cripple us here in St. Bernard Parish. But it goes much farther than St. Bernard. It ain't gonna be good."

One week in, the slick already covers 1,800 square miles. Larger than Rhode Island.

One week in, BP plans to lower a large dome. A dome to capture the oil. To capture the oil, then pump it through pipes. To pump it through pipes to a vessel on the surface. First engineers have to design it. Then workers have to make it. Could have used Better Planning. Bereft of Preparedness.

We will now play a game called BP Says. BP says it'll do this; BP says it'll try that; BP says it has ideas; BP says it needs a month—. Perhaps its leaders don't want to raise any expectations. BP spokesman, he says: "That kind of dome-pump-pipe-ship system has been used in shallow water." BP spokesman, he says: "It has never been deployed at five thousand feet of water, so we have to be careful."

Careful.

Here's how careful: "From the air, the oil spill reached as far as the eye could see. There was little evidence of a major cleanup, with only a handful of vessels near the site of the leak," writes the Associated Press on April 27.

Louisiana governor Bobby Jindal asks the Coast Guard to use containment booms, "which," he notes, "float like a string of fat sausage links to hold back oil until it can be skimmed off the surface." Oh, the innocent optimism of those early days.

The letting go of optimism. One week in, we learn the phrase "relief well." The pressure driving oil and gas out of the well may overcome any attempt to stop the blowout by pumping material in from the top. That's like trying to stuff pudding into a fire hose. So BP might decide to drill a whole other well. More likely, two. A parallel well could let workers pump material into the original well near the bottom, increasing the chances that they can clog and seal the well. No guarantees. BP says it will be a $100 million effort. We are so grateful for their generosity. And we're told that the relief wells could take an inconceivable two months. Actually, it will require twice that time. Meanwhile, they're going to try some other things. They don't know what.

The newly arriving pessimism. "You will have off-flavors that would be a concern," the oyster farmer says.

Misreported: "If the well cannot be closed, almost 100,000 barrels of oil, or 4.2 million gallons, could spill into the Gulf before crews can drill a relief well to alleviate the pressure." It's noted that the *Exxon Valdez*, the United States' worst oil spill to date, leaked 11 million gallons into the waters and onto the shores of Alaska's Prince William Sound in 1989. Compared to that, the forecast 4.2 million is meager. But the Associated Press is guessing one hundred days based on us being told three to four months, at 1,000 barrels a day.

Soon the estimate of the flow below gets quintupled. On April 28, Coast Guard rear admiral Mary Landry reports that federal experts have concluded that 5,000 barrels a day are leaking. BP had estimated—or at least said—only 1,000. Landry says BP officials are "doing their best."

If that's BP's best, well, maybe it is.

As if New Orleans doesn't have enough problems. As if shrimpers and fishermen, already staggered, need an oil spill that will finish them. Five *thousand* barrels a day. That's 200 barrels an hour. When will it end? Now I'm hearing, "It could eclipse the *Exxon Valdez*." And if the oil reaches shore—.

I read, "A BP executive on Thursday agreed with a U.S. government estimate that up to 5,000 barrels a day of crude could be spilling into the ocean."

They *agreed?* Or got busted?

Mary Landry had said there was "ample time to protect sensitive areas and prepare for cleanup should the oil impact this area."

She makes the mobilization feel like a Kabuki dance.

And now that government scientists are saying "multiply by five," the reassuring lullabies evaporate. There is snarling. Fingers pointed. Blame to go round. Now Landry warns that if not stopped, the spill could end up being among the worst in U.S. history. Now it's *mature* to say it's catastrophic. Now President Barack Obama says he'll deploy "every single available resource." He orders his disaster people, his environmental people, to the Gulf in person.

Why'd it take so long?

Five thousand barrels a day, roughly 200,000 gallons. A few of those gallons—more than anyone would like—begin lightly touching shore here and there in Louisiana. We begin hearing the name Billy Nungesser. The president of a place called Plaquemines Parish, Nungesser says a sheen of oil has reached the fragile wetlands of South Pass. (South Pass is one of the artificial mouths of the Mississippi River that was created by destroying fragile wetlands.) We begin hearing of a place called the Chandeleur Islands. On those isles of fabled fishing and abundant birds, a light sheen of oil plays touch-and-go.

Officials immediately halt fishing in a large area. A group of shrimpers sues for damages. For all the people affected by fisheries closures—everyone from people who grow oysters to people who build weekend homes—the fishing closures answer the question "How bad will this get?" by pushing the needle all the way. At least for the foreseeable future, it has just gotten as bad as it could get. They are suddenly completely out of business. People whose lives are about getting up and going to work are no longer going to work. A certain shock settles in.

Trying to regain footing on the moral high ground, a White House spokesman acknowledges that the administration might revisit the president's announcement on expanding offshore drilling.

Rear Admiral Mary Landry has repeatedly asserted that BP is the responsible party and will shoulder the costs and organizational duties associated with the cleanup effort while the Coast Guard monitors and approves things. But BP's interests are not fully in line with the public's interests. The public wants to know; wants to see.

We don't normally put the criminal in charge of the crime scene. The perpetrator's interests are different from the victim's. Certainly, as far as damages and people out of work, they'd Better Pay. But many people (including me) think the government should push BP out of the way on everything else. The company had a permit to drill. Not a permit to spill. They're on our property now. BP should be made to focus exclusively on stopping the eruption. All the big oil companies should now be convened in a war room for their best expertise. Our government should direct the efforts on everything else.

But the government keeps deferring to BP. Obama does not federalize the situation. Maybe he's afraid this will become—as people have been wondering out loud—"Obama's Katrina." Maybe he, too, wants to reserve all the blame for BP.

Today's vocabulary word: "dispersants." Use it in a sentence: "Most oil floats, but it dissolves into the sea if you apply *dispersants*." By April 30, BP has begun sending dispersants down a mile-long tube from a ship. Releasing such chemicals on the deep seafloor—rather than spraying them on surface oil—has never been done before. It's a secondary toxic leak, this one intentional, sent from above to meet the oil coming from below.

If you're BP, if part of your liability will depend on *how much oil*, it's in your financial interest to do everything you can to: (1) say you think it's leaking at a much smaller rate than it is and (2) hide as much of it as possible and (3) in as many ways as possible, try to prevent people from seeing the parts you can't hide. If you're BP, will you let people see and measure a sea of Billowing Petroleum? Not if you can avoid it.

Dispersed oil stays in the ocean. Because it dissolves into the sea, it's impossible to see or measure. Like a cake "hides" a rotten egg mixed into the batter, dispersants hide the oil. It's still a rotten egg, but now you can't retrieve it.

Dispersants are basically like dishwashing detergents, which dissolve oil and grease. And by dissolving oil concentrated at the surface, they ensure pollution of water that is home to fish larvae, fish eggs, and plankton. At certain concentrations, dispersants are toxic to those fish larvae, fish eggs, and plankton.

And what do the dolphins think? Does it burn their eyes? What is its smell to them now? What taste is conferred to their big, aware brain? Capacity for reason, ability to express; they alone in the sea can use syntax, the building blocks of language. Do they have words for this? I don't.

A week in, as the sheer unstoppable quantity of oil breaches the confines of our ability to imagine what to do, the idea of skimming up most of the oil gives way to a desire to get rid of the oil, no matter how. Landry talks about burning it off the sea. Lighting the sea on fire. In 1969, the Cuyahoga River in Ohio caught fire. So shocked was America to see its *waters* on fire that the incident—along with an oil leak off Santa Barbara, California, the same year—helped precipitate the explosion of environmental laws that Republican president Richard Nixon signed in the early 1970s.

(Note to the young: If you wonder why we need a Clean Water Act—a reasonable question, since you're lucky enough not to have seen America's waterways as they were before—consider that the Cuyahoga River had ignited about ten prior times in the last century, and also consider this description from *Time* magazine, August 1, 1969:

> Some river! Chocolate-brown, oily, bubbling with subsurface gases, it oozes rather than flows. "Anyone who falls into the Cuyahoga does not drown," Cleveland's citizens joke grimly. "He decays." . . . The Federal Water Pollution Control Administration dryly notes: "The lower Cuyahoga has no visible signs of life, not even low forms such as leeches and sludge worms that usually thrive on wastes." It is also—literally—a fire hazard.

Now Landry explains about burning this oil off the surface of the Gulf. She uses the term "controlled burns," and says they would be done far from shore. As if nothing that breathes air lives out in the Gulf of Mexico. No turtles, say. No whales. She says that crews would make sure marine life and people were protected and that work on other oil rigs would not be interrupted. But can they protect sea turtles? Can they avoid interrupting dolphins and whales?

Two decades after the *Exxon Valdez* ran aground, its oil can still be found under rocks along Prince William Sound. Scratch and sniff.

There's that terror here—that it will never be back to normal. Or that in the years it takes, communities will die and families disintegrate. But others begin saying maybe not. This isn't Alaska crude. Gulf crude is "sweet crude"—gotta love these funky terms—while the *Exxon Valdez* disgorged heavy crude. This isn't Prince William Sound. It's hot here. What's different is different.

"You have warm temperatures, strong sunlight, microbial action. It will degrade a lot faster," says Ronald S. Tjeerdema, a University of California toxicologist who studies the effects of oil on aquatic systems. "Eventually, things will return to normal."

But how long is eventually? One summer? A year? Five? Twenty? People know that before it returns to normal they are hurtling toward a time like no other, when everything they know and love is suddenly at risk.

We're all aware that there'll be oiled birds, dead turtles, the like. Some people ask, How bad? Most simply believe it will be real bad. Now we wait.

The Gulf of Mexico had gotten its own personal warning call in June 1979, when the Mexican drilling rig Ixtoc I blew out in about 160 feet of water. In the nine months required to stop the flow, it released 140 million gallons of crude oil into the Gulf, the world's largest accidental release up to then. It killed hundreds of millions of crabs on Mexican beaches and 80 percent of the invertebrates on Texas beaches. But other than that, its effects were poorly studied.

And since then, there's been no new preparedness. No new techniques. No prefabricated deep-sea oil-capture technology.

What's new is the depth of this well. A blowout this deep is new. But we shouldn't be so surprised.

DÉJÀ VU, TO NAME BUT A FEW

March 1967. The supertanker *Torrey Canyon,* whose captain took a shortcut to beat the tide, strands in the Scilly Isles, spilling 35 million gallons of Kuwaiti oil. People watch in horror as oil kills about 25,000 seabirds and coats the shores of Cornwall and Brittany. There is no containment or cleanup equipment, no capacity to aid wildlife or even assess the spill. It's the world's first massive marine oil-pollution event.

January 28, 1969. A blowout at a drilling rig six miles off Santa Barbara, California, sends 4 million gallons of crude oil into the Santa Barbara Channel, where it fouls beaches in the Channel Islands and exclusive parts of the mainland coast. Residents and the country are horrified by the sight of thousands of dying seabirds, elephant seals, sea lions, and other wildlife. The disaster creates significant awareness of the vulnerability of nature, as well as an impetus for the major environmental legislation that followed in the 1970s. It remains the third-largest U.S. marine spill.

September 1969. A tugboat towing the oil barge *Florida* from Rhode Island to a power plant on the Cape Cod Canal snaps its towline. The barge runs aground on boulders, spilling 180,000 gallons at West Falmouth and killing lobsters, scallops, clams, crabs, marine worms, and fish such as tomcod and scup. Though relatively small compared to some other spills, this is among the best studied. Four decades later, oil still forms a visible layer four inches below the marsh surface. Burrowing fiddler crabs still try to avoid the oil, and react slowly to startling motions such as a predator might make.

December 15, 1976. The *Argo Merchant*, which has been involved in more than a dozen incidents, including a collision and two groundings, runs aground again, this time off Nantucket, Massachusetts, and sinks. Its 7.7 million gallons of spilled oil cause a slick 100 miles long, 50 miles wide. The ship's résumé earns it a place in Stephen Pile's 1979 *Book of Heroic Failures*, "written in celebration of human inadequacy in all its forms."

March 1978. Oil again spoils the shores of Brittany, after the *Amoco Cadiz* loses its steering, snaps the towing cables of a rescue tug, and spills 69 million gallons of Iranian crude, which besmirches 400 miles of the French coast and destroys thousands of acres of oyster beds and thousands of fish, shellfish, and seabirds. Fish starve as their prey die, and fish populations remain depressed for two years. Where French authorities scrape oiled marshes, most never come back. Where they leave marshes alone, the ecosystems recover.

June 3, 1979. Ixtoc I, a Mexican drilling rig in about 160 feet of water, blows out and blows up, igniting an explosion and a fireball. The rig sinks. By the time two parallel relief wells plug it with 3,000 sacks of cement, it has hurled 140 million gallons of crude oil into the Gulf of Mexico, the world's largest petroleum accident up until that time. Oil eventually decimates mollusks and intertidal vertebrates along the Texas coast. An American graduate student and a Mexican biologist airlift 10,000 endangered sea turtle hatchlings to an oil-free part of the Gulf.

Late 1970s. President Jimmy Carter envisions a future array of energy sources and represents this vision with solar panels on the White House roof. After Carter's one-term presidency, Ronald Reagan has the solar panels dismantled and junked. By the end of 1985, when Reagan's administration and Congress have allowed tax credits for solar homes to lapse, the dream of a solar era has faded. Solar water heating has gone from a billion-dollar industry to peanuts overnight; thousands of sun-minded businesses have gone bankrupt. "It died. It's dead," says one solar-energy businessman of the time. "First the money dried up, then the spirit dried up."

1988. Occidental Petroleum's North Sea production platform Piper Alpha produces about a tenth of all North Sea oil and gas. When it explodes on July 6, it kills 167 men.

March 1989. The 987-foot supertanker *Exxon Valdez* has just been loaded with 50 million gallons of crude oil piped across the vastness of Alaska from the North Slope to a terminal near the tiny village of Valdez, on Prince William Sound. After altering course to avoid ice, the crew fails to get the ship to respond to efforts to resume course. Bligh Reef opens the vessel's belly, spilling at least 11 million gallons, which eventually reach beaches 400 miles—equivalent to the distance from Connecticut to North Carolina—from the grounded ship, fouling wilderness shores and killing extraordinary numbers and kinds of wildlife, from barnacles to salmon to killer whales, including roughly 2,000 sea otters and a quarter million seabirds. Hundreds of harbor seals die of disorientation and from brain lesions caused by inhaling toxic fumes. In many ways, the *Valdez* spill remains the world's most devastating and traumatic.

February 1996. The Norwegian-owned, Liberian-flagged, Russian-crewed *Sea Empress* grounds on the coast of Wales, spilling about 24 million gallons near densely populated seabird nesting rookeries, likely killing over 50,000 birds.

March 23, 2005. An explosion at BP's Texas refinery kills fifteen people. Determined: "willful negligence." BP's $108 million in fines fail to return any of the workers to their family dinner tables but stand as the United States' highest-ever workplace safety fines. Those record-setting fines represent less than 2 percent of BP's $6 billion profits for the first three months of 2010.

2005. Great Britain's Health and Safety Executive issues a warning about a Transocean-owned rig leased by BP in the North Sea, saying the rig's remote blowout-preventer control panel had not been "maintained in an efficient state, efficient working order and in good repair." The office also accuses Transocean of bullying and intimidating its North Sea staff.

2006. BP spills 200,000 gallons of oil in Prudhoe Bay. The North Slope's biggest spill, it is caused by corrosion in poorly maintained equipment. BP is eventually ordered to pay $20 million in fines.

September 14, 2007. In a letter, the Fish and Wildlife Service agrees with the Minerals Management Service's conclusion that deepwater drilling in the Gulf of Mexico poses no significant risk to endangered species like the brown pelican and the Kemp's ridley sea turtle. The agencies consider only spills totaling 1,000 to 15,000 barrels. And they say such spills would have less than a one-in-three risk of oiling the critical habitat for either of the endangered species.

October 2009. An environmental assessment for an area of the Gulf covering the site of the Deepwater Horizon blowout (Lease Sale 213) says, "The effect of proposed Lease Sale 213-related oil spills on fish resources and commercial fishing is expected to cause less than a 1 percent decrease in standing stocks of any population, commercial fishing efforts, landings, or value of those landings." It concludes thus: "There would be very little impact on commercial fishing."

November 2009. A BP pipeline ruptures in Alaska, releasing about 46,000 gallons of oily gunk onto the tundra.

January 2010. In a letter to the president of BP Exploration (Alaska), Congressmen Henry Waxman and Bart Stupak refer to "a number of personnel incidents involving serious injury or death" and question whether proposed BP budget cuts might threaten the company's ability to maintain safe operations.

March 4, 2010. President Obama lays out a Gulf Coast plan, "to deal with the catastrophic dangers of rising sea levels, hurricanes, and erosion, and invest in restoring barrier islands and wetlands in Mississippi and Louisiana." The White House adds, "Unless we stem the rapid rate of loss, Gulf ecosystems and the services they provide will collapse." And the Department of the Interior adds, "Finally, a president that has said we are going to take charge of this." This is not about concern over oil leaks. The concern: the Mississippi River Delta has been slowly falling apart for more than half a century. Causes: levees and flood controls built since the 1930s also starve marshes of the sediments and nutrients that created and maintained them; 14 major ship channels have been gouged through the wetlands to inland ports to facilitate Mississippi River commerce; and countless canals and channels have been dug by oil companies for boats, pipelines, and oil-rig servicing.

March 25, 2010. Three U.S. senators—Massachussetts Democrat John Kerry, South Carolina Republican Lindsey Graham, and Connecticut switch-hitter Joe Lieberman—are trying to draft bipartisan climate-change legislation capable of getting sixty votes to overcome a filibuster. It's the year of filibusterphobia. Graham's main concern is that depending on foreign oil is a security threat. The environment isn't uppermost in his mind. Considering these things, their bill includes environmental compromises so big, oil rigs could sink in them. So those three senators get a letter from ten Democrat senators. The ten warn that expanded offshore drilling could put their states at risk from oil spills, threatening fisheries, tourism, and the coast—a "national treasure that needs to be protected for generations to come." Senator Bill Nelson of Florida says the letter means "in a nutshell: No oil rigs off protected coastal states." Florida's beaches are its white gold. The senators' letter says the best way to lower oil costs is through energy efficiency and conservation.

April 4, 2010. The 700-plus-foot Chinese-owned coal carrier *Shen Neng 1* slams into Australia's Great Barrier Reef at full speed. The impact ruptures the vessel's fuel tanks. It's carrying 300,000 gallons of heavy fuel oil to run its engines while hauling 65,000 tons of coal.

First week of April 2010. The Government Accountability Office reports to Congress that the federal Minerals Management Service's Alaska office hasn't developed any guidelines for determining whether proposed developments comply with federal law. The GAO complains of "absence of a process."

April 14–15, 2010. The Interior Department's Minerals Management Service approves a series of permit changes that will help BP speedily conclude its over-budget drilling operation being conducted on Transocean's Deepwater Horizon drilling platform. BP has already secured a "categorical exclusion" from environmental review under the National Environmental Policy Act.

2010. Another BP platform, called Atlantis, has been operating with major apparent safety irregularities. One man who called attention to them has been laid off. He is suing BP. He says, "I've never seen this kind of attitude, where safety doesn't seem to matter and when you complain of a problem and try to fix it, you're just criticized and pushed aside." In 2008 a manager wrote an e-mail warning of "potential catastrophic operator errors" and said that operating the rig while "hundreds if not thousands" of critical engineering drawings for its operation were never finalized is "fundamentally wrong." The Atlantis well is capable of pumping a lot of oil, estimated at 800,000 barrels per day when fully operational. Says the man who's been laid off, "If something happens there, it will make the Deepwater Horizon look like a bubble."

PART TWO

A SEASON OF ANGUISH

MAYDAY

Perhaps no pause has ever been as pregnant. Before May Day, hotel owners, fishermen, and restaurateurs—and most of the rest of the country—begin an anxious watch of satellite images and maps of how the oil is spreading, and models of how the oil could spread. The mere thought. The fear that it could all be lost. The fun, the wildlife—and the revenue they represent—ruined.

It begins looking like it could wash ashore heavily within days, coating fragile wetlands, ruining the Gulf's famous oyster beds, and redacting the whiteness from heretofore dazzling beaches. In fact, our minds spoil those beaches before any oil reaches them. The siege mentality seizes up our thinking.

So by May Day, a certain passion play has been written: the greed of men, the sacrifice of life, utter unpreparedness compounded by inability to respond, and a blanketing dread.

"I am frightened for the country," says the assistant chief of the National Ocean Service, David Kennedy. "This is a very, very big thing, and the efforts that are going to be required to do anything about it, especially if it continues on, are just mind-boggling." And Florida's Governor Charlie Crist wonders if anyone, really, could be doing enough in this situation. "It appears to me," he said, "that this is probably much bigger than we can fathom."

A New Orleans businessman who loves the bayou and loves his fishing tells me, "We got hit with the country's worst natural disaster; now we're getting hit with the worst man-made disaster. We were finally getting past the Katrina aftermath and the stigma, with the Saints winning the Super Bowl, and many businesses finding the stimulus dollars well spent." He tells me, "We just had this great glow going. And now it's just starting to dawn on folks what this oil might mean. It's the rest of *my* lifetime, anyway."

He says that people are "hugely frightened" of both the disaster itself and the media's ability to exacerbate the situation.

"This is the worst possible thing that could happen to the Mississippi Gulf Coast," says a man in the tourism business. "It could kill family tourism. That's our livelihood."

The person I'm staying with overnight in New Orleans says, "Everyone's deeply vested in the enjoyment and all the livelihoods and businesses that have been made. This is an outdoor lover's paradise. *Everything* revolves around seafood, wild food, the marshes, the coast. I mean, people are *very* scared that it will change a way of life. You've got to get outside this city, go down to those areas. You've got to feel that."

And so, I do.

I am now bouncing around the southern Louisiana delta region, through the mixture of poverty and affluence (mostly poverty) and the many visible reminders of Katrina. People who'd been rebounding post-Katrina now feel truly scared that their economy and their future are ruined forever. It's quietly horrifying.

Seeing families in small boats, local people crabbing in the canals, and the enormous pride with which people rebuilt waterside fishing retreats that Katrina had wholly swept away, I feel many connections between the people here and my own loves.

For all its self-image as the laid-back land of the Big Easy, the Gulf Coast of Louisiana is a brutal and brutalized place. The work hard, the options few, the stakes high, the damage and the scars deep.

Docked near Shell Beach lies a red-hulled trawler named *Blessed Assurance.* These communities could always depend on the marshes and waters and fishes and shellfish, no matter what. But now, that blessed assurance is suspended indefinitely. Human dignity is its own justification, and many people here, citizens of our imperfectly united country, feel that they may have just lost theirs.

"I'd hate to think that the hurricane didn't kill me and an oil spill did," says the sixty-eight-year-old survivor who'd rebuilt his century-old bait-and-fuel business from scratch after Katrina utterly destroyed it. Says his forty-one-year-old son, "This marsh is going to be like a big sponge, soaking up the oil. It's going to be bad." Captain Doogie Robin, eighty-four, oysterman, says, "Katrina really hit us hard. And this here, I think this is going to finish us now. I think this will wipe us off the map." A seafood dealer in Buras, Louisiana, says, "When you kill that food chain, nothing's going to come back to this area." A woman in Hopedale says, "If it gets in the marsh, it could be a year, it could be two years, it could be ten years." She adds, "It all depends on what happens."

What happens: By the first week of May, fishing is suddenly a thing of the past in 6,800 square miles of federal waters. More people whose lives depend on fishing realize they're out of work. Just like that. And the effects ripple further. A restaurant owner catering to fishermen worries, "How is a fisherman going to be able to afford to eat if he doesn't have a job?"

Meanwhile, the spokesman for the Louisiana Seafood Promotion and Marketing Board is working to—y'know—promote Louisiana seafood, assuring the public that Louisiana seafood remains available and safe. In 2008, the federal government had estimated the annual volume and value of Gulf-caught fish and shellfish at 1.3 billion pounds, worth $661 million. "We should be fine unless this thing gets totally out of control," the seafood promoter says.

"What BP's doing is throwing absolutely everything we can at this," says a senior vice president for the corporation.

Does that sound like a response plan or like this thing is totally out of control?

On May 2, BP begins drilling the first of those two relief wells. I say "BP begins drilling," but actually BP doesn't actually own any drilling equipment or *do* any drilling. As always, it simply hires someone else. It's just the Big Payer. Incredibly, these relief wells are planned to go straight down to around a mile below the seafloor (two miles from the surface), then *angle* in toward the blowing-out well, whereupon they will hit it 13,000 feet below the seafloor—where it is seven inches in diameter. They're capable of doing this—but they can't stop a leaking pipe lying on the seabed.

BP says it will spend a week making a 74-ton, concrete-and-metal box to put over the leak. "It's probably easier to fly in space than do some of this," says a BP spokesman.

Florida's Governor Charlie Crist says, "It's not a spill, it's a flow. Envision sort of an underground volcano of oil and it keeps spewing."

Twenty or so dead sea turtles wash ashore along thirty miles in Mississippi. None with oil on them, but the number's unusual. They could have eaten oil blobs, a frequent health problem for sea turtles. Or been overcome by fumes. Or drowned in shrimp nets. Fingers get pointed each way. Shrimpers say no way, impossible; they're required to use turtle-escape devices in their nets. The suspicious say panicked shrimpers tied the turtle-escape flaps shut to get every last shrimp. I'm going with the shrimpers on this one. I think it was the oil. A few years ago, off South Carolina, where there had not been any spills noted, veterinarians showed me how ingesting oil suppresses turtles' white blood cell counts, their immune system, and makes them unhealthy.

May 7. Along with the oil comes a new vocabulary of silly names for half-baked ideas. Today's password: "dome." Use it in a sentence: "BP has constructed a four-story containment dome intended to control and capture the largest of the leaks (yes, the pipe is leaking elsewhere, too)." But even as the dome was lowered, crews discovered that the opening was becoming clogged by an icy mix of gas and water—why didn't the engineers foresee this?—so they set the 74-ton steel contraption down on the seabed 650 feet away from the leak, "as officials decide how to proceed." New definition of "decide": to scratch one's head in wonder and confusion.

Foreseeing vast costs for cleanup and damage, investors Begin Pummeling, wiping about $30 billion off BP's value in the first two weeks of the blowout. Outrage spreads. Momentum billows for lifting the recklessness-inducing $75 million cap on oil firms' damage liability to a more realistic $10 billion.

It doesn't take long to connect the most obvious dots: The *Economist* argues that "America's distorted energy markets, not just its coastline, need cleaning up." America needs a real energy policy. We need more diversified, cleaner, more competitive energy.

Easier said.

Meanwhile the *New York Times'* Thomas Friedman writes that there is only one meaningful response to the horrific oil spill in the Gulf of Mexico and that is for Congress to pass an energy bill that will create an American clean-energy infrastructure and set our country on a real, long-term path to ending our oil addiction. The obvious beneficiaries: our environment, our national security, our economic security, innovators, entrepreneurs. "We have to stop messing around," writes Friedman, "with idiotic 'drill, baby, drill' nostrums, feel-good Earth Day concerts and the paralyzing notion that the American people are not prepared to do anything serious to change our energy mix." He says this oil spill is like the subprime mortgage mess, a wake-up call and an opportunity to galvanize "radical change that overcomes the powerful lobbies and vested interests that want to keep us addicted to oil."

It all sounds so obvious, and too familiar. And that's the problem: the things that should have us on fire demanding change somehow fail to rouse we, the people, to the passion that could free us from our dependence on our pushers.

Senator John Kerry, one of the first sponsors of the Senate's energy bill, back before the blowout, seems to share such sentiments. He believes America is confronting three interrelated crises: an energy security crisis, a climate crisis, and an economic crisis. He says our best response to all three "is a bold, comprehensive bill that accelerates green innovation and creates millions of new jobs as we develop and produce the next generation of renewable power sources, alternative fuels and energy-efficient cars, homes and workplaces."

But ironically, because some Republican governors—such as Florida's Charlie Crist and California's Arnold Schwarzenegger—are withdrawing support for expanding drilling off their coasts, Republican support for an energy bill—wafer thin at its apogee—is dissolving faster than dispersant-drenched oil. Obama paid the price of reaching for that support by opening new areas, but the blowout has coastal politicians shrinking back in horror. The bill dies. Most proponents say the bill in Congress had been decorated with so many gifts and compromises that it had become a bad bill anyway. That's an apt balm. It was like the bill had eaten so much fat that it collapsed of arterial blockage.

Even though it's a matter of physics that carbon dioxide makes the planet warmer, and even though, because of burning oil and coal, our atmosphere now contains a third more carbon dioxide (and climbing) than it did at the start of the Industrial Revolution, many people just won't believe we have a big problem. People have a lot of kooky notions, but many who are disconnected from reality on this issue are running or aspiring to run the government of the United States of America.

Ron Johnson, a Republican candidate for the U.S. Senate from the great dairy state of Wisconsin, doesn't "believe" that carbon dioxide is causing climate change, "not by any stretch of the imagination." With his head still in his dairy air, he's going to Washington. In a few months, Johnson, steeped in tea party support, will beat Democratic incumbent Russ Feingold. "I think it's far more likely," Johnson opines of the causes for global warming, "that it's just sunspot activity."

When U.S. senators are offering the same explanation for climate change that were used to explain UFOs in the 1950s—watch out.

Here is a *real* conservative, one I admire: Congressman Bob Inglis, Republican of South Carolina:

> As a Republican, I believe that we should be talking about conservation, because that's our heritage. If you go back to Teddy Roosevelt, that's who we are. And after all, there are very few letters different between conservatism and conservation. People asked me if I believe in climate change. And I tell them, no, I don't believe in climate change. It's not big enough to be a matter of faith. My faith informs my reaction to the data. But the data shows that there is climate change and that it stands to reason that it is in part human caused. And so, therefore, as responsible moral agents, we should act as stewards. Unfortunately, a clear majority of the Republican conference does not accept human causation in climate change. It's definitely not within the orthodoxy of conservatism as presented by, you know, Sarah Palin and folks like her. So you don't want to stand against that. And the result is that some people are sort of cowed into silence. . . .
>
> It's really about national security. We are dependent for oil on the region of the world that doesn't like us very much. We need to change the game there. It's also about free enterprise, letting the free enterprise system actually solve this problem. Right now, the reason that free enterprise can't solve the problem is that petroleum and coal have freebies. Accountability, by the way, is a very bedrock conservative concept, even a biblical concept. We insist on accountability. So do we want to persist in a situation where we need to land the Marines in order to make sure that we have, in the future, the access to this petroleum that we must have? Or do we want to use the strength of America, the free enterprise system and the innovativeness of entrepreneurs and investors here to break that dependence? So our choice is, do we play to our strengths? Or do we continue to play to our weakness, which is playing the oil game?

Anyone expecting thunderous applause for such a rousing call to patriotism isn't paying attention. Asked what happens to Republicans who respond positively to science, Inglis says, "People look at you like you grew an extra head or something. You're definitely seen as some kind of oddball, and perhaps even a heretic."

Congressman Inglis lost his bid for reelection in the 2010 *primaries.* Too "moderate."

The fact that politicians, media talking heads, and too much of the electorate lost the ability to differentiate between science and ideology is one of the causes of America's decline. But if Americans don't understand science, is it really fair to blame Big Oil?

The government's main office for gathering all the scientific data on climate change and informing the U.S. Congress, more than a dozen federal agencies, and the American people is called the Global Change Research Program. Rick Piltz was a senior officer there from 1995 to 2005. Soon after George W. Bush took over the White House after losing the "popular vote"—which in other countries is called the "election"—Piltz was putting together a major report for Congress. "We were told to delete the pages that summarized the most recent IPCC report and the material about the National Assessment of climate change impacts that had just come out," he recalled in 2010. The IPCC is the international scientific body that collects and assesses all the climate research from around the world. The National Assessment was a similar report covering research by U.S. scientists. They'd both concluded that climate change was happening and that human activity was accelerating it.

But the Bush White House put its fingers in its ears and sang, "La-di-da." Piltz says the experts had made "pretty clear and compelling statements. And to say that you didn't believe it was to say that you did not want to go along with the preponderance of scientific evidence." Until he left, four years later, from almost every report Piltz and his team compiled, the White House deleted references to climate change or carbon emissions. Many of those deletions were made by a guy named Philip Cooney. He was chief of staff for the Bush White

House's Council on Environmental Quality. His previous job had been as a lawyer and lobbyist for the American Petroleum Institute, and his next job was with ExxonMobil.

And when we so crucially need campaign finance reform and publicly funded elections to get the money out of politics, in the 2010 Citizens United case, the Supreme Court overturned a century of precedent, and effectively destroyed the McCain-Feingold campaign finance reform legislation. McCain-Feingold banned the broadcast or transmission of "electioneering communications" paid for by corporations or labor unions from their general funds in the thirty days before a presidential primary and in the sixty days before the general elections. By doing away with that, the Supreme Court opened the floodgates to corporate election-distorting money and made it easier for the sources of that money to remain anonymous. All five of the Court's "conservatives" joined together to overturn a sixty-three-year-old ban on corporate money in federal elections and twenty-year-old and seven-year-old precedents affirming the validity of such corporate electioneering bans. They ignored the protests of their four more moderate (actually conservative, in other words) dissenting justices. Writing ninety pages for the dissenters, Justice John Paul Stevens noted: "Today's decision is backwards in many senses. The Court's opinion is a rejection of the common sense of the American people, who have recognized a need to prevent corporations from undermining self government since the founding, and who have fought against the distinctive corrupting potential of corporate electioneering since the days of Theodore Roosevelt. It is a strange time to repudiate that common sense. While American democracy is imperfect, few outside the majority of this Court would have thought its flaws included a dearth of corporate money in politics."

President Obama called the decision "a major victory for big oil, Wall Street banks, health insurance companies and the other powerful interests that marshal their power every day in Washington to drown out the voices of everyday Americans."

This "Joke of the Week" arrives in my in-box: "The economy is so bad, Exxon-Mobil just laid off 25 congressmen."

❦

I hop onto a small boat that's already pulling away from the dock in Shell Beach, Louisiana. The captain, Casey Kieff, was a fishing guide until the oil blowout caused an indefinite fishing closure. Now he's taking some photographers out to have a look at Breton Sound. On board is a reporter from Reuters, a crew from Russian state television, a photographer working for Getty Images, and several others. The captain names a fair price and I jump on, too, wondering if this is his last week of work—and what price could really be fair to him. For decades he's been guiding people who want to sport-fish for speckled sea trout and redfish. No matter how bad the oil gets, the media attention will wane. Guiding reporters won't be a new career. So he's got a lot on his mind.

We go to Hopedale to pick up a couple of other people. The cops have the road into here closed, but by the grace of boats we have free range of the place. The waterfront is bustling with trucks carrying miles and miles of boom, all kinds of boats getting into the act for a day's uneasy pay. The National Guard, wildlife enforcement people—all kinds of busyness intent on oil-containment plans that are at worst futile and at best high-risk. There seems both a lot of organization and a lot of confusion.

In Kieff's overpowered outboard, we blast through a sliver of the astonishingly vast and intricate wetlands of Louisiana.

The first thing that really impresses me is the immensity of Louisiana's marshes and coast. Marshes as far as you can see, to all points of the compass. Bewildering mazes of channels.

How could this whole coast be protected? And if oil comes, it could never be cleaned by people. It's a wet, grassy sponge from horizon to horizon.

The skeletons of oak forests stand starkly on marsh islands now subsiding, too low and too inundated to sustain trees. They're dying back, and the marshes themselves are eroding away.

But still, there are miles and miles of marshes before one gets to the open waters of Breton Sound. As we head toward the Gulf, I notice a few bottlenose dolphins rolling as they snatch air in the mud-murked channels.

When we reach open water, the horizon is dotted with gas rigs and a few boats that have piled their decks with booms.

After a long, pounding ride through a stiff chop, our captain steers us to one of the inner islands, Freemason, a barely emergent ridge of sand and shell about a mile long. The innermost of the Chandeleur chain, the island is maybe a mile long. (Katrina dissolved several miles of these islands.)

The island isn't well known to people. But it is to birds. Brown pelicans, laughing gulls, herring gulls, black skimmers, and least, Sandwich, and royal terns rest by the dozens. They're joined by a few itinerant ruddy turnstones, sanderlings, and black-bellied plovers migrating toward their Arctic nesting grounds.

In some ways, a catastrophe of this magnitude could not have happened in a worse place. Or at a worse time of year.

The Gulf is a large region, but its natural importance is even more outsized, disproportionate to its area. The Gulf is the hourglass pinch point for millions of migrating creatures that funnel in, and then fan out of it to populate an enormous area of the hemisphere's continents and coasts. Anything that affects living things inside the Gulf affects living things far outside it.

In the Gulf in May, with the oil gushing, are loons, gannets, various kinds of herons and others that have spent the winter here but will soon leave to migrate north. Depending on the species, they'll breed all along the coast from the southern states to as far north as the Maritimes and lakes across much of Canada. Some of the longest-distance migrants on Earth are various sandpipers, plovers, and other shorebirds, many of which winter as far south as Patagonia and breed as far north as the high Arctic. Perhaps a million cross the Gulf in May, and when they reach the U.S. coastline, they must stop to rest and feed. Problems with habitat and food supply have reduced many of their populations 50 to 80 percent in the last twenty years. And now this.

Oil is just starting to smudge some of the birds. Even among those that do not get heavily oiled, some will not make it. The birds' energy budgets will not bear the cost of feathers sticking and functioning inefficiently, and many such birds will likely drop out on their way north. Migrating peregrine falcons traveling north from South American wintering areas, destined for nesting sites as far as Greenland, are also crossing the Gulf's marshes. Preferentially picking off birds whose flight seems compromised, falcons could end up getting disproportional doses of oil.

Certain animals that normally inhabit the open Atlantic travel to the Gulf to breed. The world's most endangered sea turtle, the Kemp's ridley, ranges throughout the western Atlantic as far north as New England. But it breeds only in the Gulf. Adults are now heading there to lay their eggs on remote beaches. So are other sea turtles. Adults are vulnerable, but hatchlings will likely have an even harder time. And whether from oil or fishing nets, disproportionate numbers of turtles continue turning up dead.

A magnificent frigatebird—that's not my description; it's the species' name: magnificent frigatebird—patrols overhead. It shares the airspace with a couple of helicopters and a C-130 military cargo plane, newcomers in its eons-old realm. And it feels the place more deeply than anyone aboard those aircraft.

Booms designed to keep away oil have been placed along one side of Freemason Island. But they don't run the whole length of even one side of the island. And on the side where they were placed, the wind and chop have already washed them ashore in places and partly buried them in sand and shell. In other words, segments of the boom barrier have already been rendered useless by a couple of days' wave action.

We'd had some news that part of the oil slick was eight miles southeast of the islands in the open Gulf. But the water is too rough for us to continue on to the main Chandeleur Islands or beyond. Indeed, some of the fishing boats carrying booms are headed back in. Over the radio they tell us they were sent toward port due to the rough water.

So far, I see oil only on the heads and bellies of a few Sandwich terns. Just a few little brown smudges; but it's unmistakably oil. I fear much worse is coming. Diving into water—that's how they eat. Oil is famously hazardous to waterbirds. It's a chronic thing: a few oiled birds are always showing up around harbors and ports, and I once saw a tropicbird in the middle of the ocean whose immaculate pearl plumage was stained with considerable oil from somewhere—probably a ship that had discharged dirty bilgewater.

People have been counting dead birds from major spills for years. In 1936, 1,400 oiled birds washed ashore near Kent, England. In 1937, a ship collision sent 6,600 oiled birds onto the coastline of California. Five thousand ducks on the East Coast in a 1942 mishap; 10,000 ducks killed by oil in the Detroit River in 1948; an estimated 150,000 eiders off Chatham, Massachusetts, after two ships collided in a 1952 winter gale; 30,000 ducks off Gotland, in the Baltic, in 1952, and 60,000 more in the same place in three successive spills over the next four years. In 1955 the wreck of the *Gerd Maersk* killed an estimated 275,000 scoters and other ducks and waterbirds near the mouth of Germany's Elbe River. Kills of several thousand birds remained quite common throughout the 1960s. As Japan industrialized, its first large oil spill, in 1965, killed large numbers of kittiwakes, shearwaters, and other birds. Between 1964 and 1967, worldwide there were 19 tanker groundings, with 17 large spills, and 238 supertanker collisions, resulting in 22 more large spills. Though thousands of birds continued to be killed in various oil-related mishaps, improved safety and environmental regulations resulted in fewer accidents and reduced wildlife deaths, until the *Exxon Valdez* set new records.

Birds that are lightly oiled, like the ones here today, often raise fewer, slower-growing chicks than normal.

Some oiled animals may eventually be rescued and cleaned. Less possible to cleanse is the anguish on the faces and in the hearts of fishing families.

Casey wants to say hey to a relative, James Kieff, who's been an oysterman for thirty-five years. "We outta work," James says disbelievingly from the deck of his forty-eight-foot boat, *Lady Jennifer*, as he motors along carrying 3,000 feet of boom to the outer edge of one of the marsh islands. "Now we workin' for BP. They don't know these waters. We do." A moment later he adds, "The stress is in the not knowing. Katrina came and went. We knew what to do. If the oil stays offshore, we could be okay. But that's a big if."

I watch him and other boats laying booms along the marshes. They work hard. I see men and women who know water, boats, and work, bending seriously to their task. And despite never having handled booms before, they do the job well.

The only problem: No one believes the booms can work. Any medium wave action will push oil over them. And despite miles and miles and miles of booms, miles and miles and miles and miles and miles and miles of marsh remain utterly naked and undefended. It's a fool's errand. (During the *Valdez* catastrophe, Exxon's CEO was apparently audiotaped saying he didn't care if the booms contained the oil; he just wanted pictures of them in the water.)

The only saving grace is that there is no oil in sight yet. And that hardly helps. Because the people have no jobs after this. As always, it's a matter of who people are at the mercy of how people are.

But for this frantic moment, with oil rigs in every direction and the oil coating everyone's mind, it's boom boom boom.

A Louisiana State University professor says the oil could entirely wipe out many kinds of fish, and notes, "We may very well lose dozens of vulnerable fish species." A toxicologist says, "We'll see dead bodies soon. Sharks, dolphins, sea turtles, whales; the impact on predators will be seen in a short time because the food web will be impacted from the bottom up." Another Louisiana State University professor says oil pushed inland by a hurricane could affect rice and sugarcane crops. A meteorologist says there's a chance that the oil could cause explosive deepening of hurricanes in the Gulf. The *Christian Science Monitor* asks whether hurricane-blown oil could make coastal towns permanently uninhabitable. The *Monitor* then quotes a Pensacola Beach solar energy salesman (bias alert) who took a hazardous materials class as saying, "In these classes, they basically tell you that swallowing even a small amount of the oil or getting some on your hands and then having a smoke could be deadly."

Well, holy cow; somebody needs to warn auto mechanics, boat owners, auto-lube attendants, heating-oil delivery people, gas station attendants, and anyone with a car or lawn mower *not to move* because they're about to *explode!*

In the Florida Keys, Miami, and beyond, people begin worrying about the "Loop Current" (vocabulary term). It flows out of the Gulf of Mexico to become the Gulf Stream, and might carry oil throughout the Keys' reefs and then up the East Coast. "Once it's in the Loop Current, that's the worst case," says a Texas A&M University oceanographer. "Then that oil could wind up along the Keys and get transported out to the Atlantic." I begin hearing worries from Long Island's Hamptons—and, indeed, as far away as *Ireland*—that in a few months, oil from the blowout will ruin beaches there. Florida senator Bill Nelson warns, "If this gusher continues for several months, it's going to get down into the Loop Current. You are talking about massive economic loss to our tourism, our beaches, to our fisheries, very possibly disruption of our military testing and training, which is in the Gulf of Mexico."

Another scientist from Texas A&M University says, "The threat to the deep-sea habitat is already a done deal, it is happening now." He adds, "If the oil settles on the bottom, it will kill the smaller organisms, like the copepods and small worms. When we lose the forage, then you have an impact on the larger fish."

Yeah, maybe. But "a done deal"? Really? How much oil would have to settle? How densely?

I'm a professional environmentalist and conservationist; I'm really angry about the recklessness that caused this, and the inanity of the response; I am deeply distressed about the potential damage to wildlife and habitats—but I find myself becoming uncomfortable with all the catastrophizing. "A done deal"—that's not very scientific. Especially for a scientist. Many scientists—and as a scientist it hurts to say this—are being a little shrill. Cool heads are not prevailing.

But it's not a time of calm. With the situation out of control, everyone wants to know what's gonna happen. Even the normally cautious are prone to overspeculating.

And then there are those never burdened by caution. Enter the crazies. Some say this was done on purpose: Obama and BP have conspired to make money from this; someday, they claim, we'll get to the bottom of how. Some believe this will merely kill the entire ocean; others think it will kill the whole world.

Something called Yowusa.com posts an article based on warnings it says are from an Italian physicist. (The site also features predictions made by Nostradamus, including his "proven" prediction that a giant tsunami will destroy New York City.) We read, "The Loop Current in the Gulf of Mexico has stalled as a consequence of the BP oil spill disaster. The effects have also begun to spread to the Gulf Stream. If natural processes cannot re-establish the stalled Loop Current, we could begin to see global crop failures as early as 2011." Yow, indeed.

The Web begins amplifying a blogger-posted article called "How BP Gulf Disaster May Have Triggered a 'World-Killing' Event." In another post the same blogger writes, "The giant oil company is now quietly preparing to test a small nuclear device in a frenzied rush against time to quell a cascading catastrophe." On the same site (whose slogan is: "Where Knowledge Rules") we find several entries under the heading "How the Ultimate BP Gulf Disaster Could Kill Millions." The first click gets me "A devastating eruption of methane gas, buried deep beneath the sea floor, could absolutely decimate the region. Even worse, this eruption could inundate the low lying coast line with a tsunami." The entry unleashes a tsunami of speculation and dire conclusion that surges across the Web.

Sparked by very legitimate fears and fanned by wild speculation, a certain simmering siege mentality, peppered with panic, begins bubbling across the region. The sky begins falling a little.

As gallop horsemen of the Apocalypse, so also ride a few doubters of disaster. "The sky is not falling. It isn't the end of the Gulf of Mexico," says the director of a Texas-based conservation group. While many are predicting a thousand ruined miles of irreplaceable wetlands and beaches, fisheries sidelined for seasons on end, fragile species shattered, a region economically crippled for years, and tongues of oil lashing beaches up to Cape Cod, others shrug.

It's still the first inning in a nine-inning game. "Right now what people are fearing has not materialized," says a retired professor and oil spill expert from Louisiana State. "People have the idea of an *Exxon Valdez*, with a gunky, smelly black tide. . . . I do not anticipate this will happen here." Others point out that the Ixtoc leak seemed largely to vanish in about three years. (Even if it did, three years would be a very long time in a fishing community that can't go fishing.)

Still others note that this isn't so bad compared to all the other bad stuff that happens to the Gulf every day: scores of refineries and chemical plants from Mexico to Mississippi pouring pollutants into the water, pollution from all the ships, the degrading marshes, the dead zone caused by all the Midwest's farm runoff flowing down the Mississippi River. It's hardly a pristine Eden.

"The Gulf is tremendously resilient," says the cool-headed director of the Texas-based conservation group. But he adds, "How long can we keep heaping these insults on the Gulf and having it bounce back? I have to say I just don't know."

No one knows. The not knowing thickens a gumbo of fear.

Meanwhile, BP executives admit to Congress behind closed doors that the leak could reach 60,000 barrels per day, sixty times what they'd been saying.

Now, about that Loop Current. There're kernels of truth in both the fear that it could take oil across the Keys reefs and into the Gulf Stream and the observation that it's "stopped." Normally, the Loop Current flows a bit like a snaking conveyor belt; it does have the potential to take the oil and move it past the Florida Keys. But that conveyor has just pinched itself off. As sometimes happens, a meander has bent itself into an enormous eddy that's just pinwheeling in the Gulf. This greatly blunts the likelihood of the oil getting into the Gulf Stream and up the East Coast. "This is the closest thing to an act of God that we've seen," says Dr. Steve Murawski of the National Oceanic and Atmospheric Administration. Well, maybe—whatever. But another way of looking at it is that the Gulf will be stuck with all the oil. And I'm about to get a pelican's-eye impression of how much oil is currently in the Gulf.

"My name's Dicky Toups," says the pilot before he starts the engine of the seaplane I'm climbing into. "But they call me Captain Coon-ass."

We overfly the emerald maze of the vast Mississippi Delta. Captain Coon-ass points, saying, "There's a big ole gator."

To the far points of view, America's greatest marshes lie dissected, bisected, and trisected, diced by long, straight artificial channels and man-angled meanders, all aids to access and shipping. For the vast multimillion-acre emerald marshes, they are death by a thousand cuts.

The sign had said, "Welcome to Louisiana—America's Wetland." Pride and prejudice. People here depend on nature or the control of nature—or both. Keep an eye on nature; it can kill you here. But people can kill the place itself.

Since the 1930s, oil and gas companies have dug about 10,000 miles of canals through the oak and cypress forests, black mangrove swamps, and green marshes. Lined up, they could go straight through Earth with a couple thousand miles to spare. The salt water they brought killed coastal forests and subjected our greatest wetlands to steady erosion. Upstream, dams and levees hold back the sediment that could have helped heal some of that erosion. Starved on one end, eaten at the other. How to kill America's wetlands. Long after this oil crisis is over, this chronic disease will continue doing far more damage than the oil.

All these Delta-slicing channels cause banks to dissolve, swapping wetlands for open water. Those channels also roll out red carpets for hurricanes. Incredibly, this has all cost Louisiana's coast about 2,300 square miles of wetlands. Marshland continues to disintegrate at a rate of about 25 square miles a year. The rise in sea level due to global warming is also helping drown watery borderlands. Oil leak or no leak, these things, all ongoing, constitute the most devastating human-made disaster that's ever hit the Gulf. Bar none.

Only slowly does the muddy Mississippi lose itself to the oceanic blue of the open Gulf, a melding of identities, a meeting of watery minds. And also the drain for sediments, agricultural fertilizers, and dead-zone-generating pollutants from the entire Midwest and most of the plains. Even before the oil blowout, this was a troubled place—a troubled place whose troubles have now escalated to a whole new level.

Two boats are tending booms around an island densely dotted with nesting pelicans. As I've noticed from the ground—but it's even more striking from up here, at 3,500 feet—most of the coast is bare of booms and undefended. Where booms have been placed along the outer beach, many have already washed up onshore, already useless.

The sea-surface breeze pattern is interrupted by a marbling of slicks. Often such a pattern is perfectly natural, so I look carefully. It's brown.

"Oil," says Captain Coon-ass.

One of those slicks has nuzzled against the shore. There's a boat there and some people are walking along the beach, inspecting a long boom that the wind has washed ashore.

The nearshore waters and beyond are dotted with drilling rigs for oil and gas, some abandoned. Like bringing coals to Newcastle, many of the rigs stand surrounded by floating oil.

Offshore, longer slicks ribbon their way out across the blue Gulf. As we follow them, the light slicks thicken with dark streaks that look from the air like wind-driven orange fingers, then like chocolate pudding. An ocean streaked with chocolate pudding.

A few miles out, the streaks grow darker still. Yet there remains far more open water than oil slick.

That changes. Blue water turned shiny purple. A bruise from a battering. The sea swollen with oil.

As the water darkens and the slicks widen, Captain Coon-ass points to a small plane below us, saying it's on a scouting run for the C-130s that will follow to spray dispersants. More chemicals on a sea of chemicals.

Yet plenty of the oil—and I mean plenty—is not dissolved. Blue water turned brown.

"This is some pretty thick stuff right here," Captain Coon-ass says. The crude is now drifting in broad bands that stretch to the horizon. "We're lookin' at twenty miles of oil right here."

We're directly over the source of the blowout. Below, two ships are drilling the relief wells that we've been told will take months. A dozen ships drift nearby, most with helicopter landing pads on them. What they're all doing, heaven knows.

A fresh breeze puts whitecaps on the nonoiled patches of the black-and-blue sea. As the C-130 comes out, we turn northeast.

We're headed toward the Chandeleurs, the line of sandy islands that have been much in the news for their at-risk bird rookeries. Soon we're over Breton Sound, where a couple of days ago, from a boat, I saw no oil.

But now there is plenty of oil, moving in between the main coast and the islands.

Out to intercept the oil is a fleet of shrimp boats towing booms from the outriggers that would normally tow their nets. The idea appears to be that they will catch the oil at the surface, the way they catch shrimp at the seafloor.

Dozens of boats tow booms through the oil, but as they do, water and oil simply flow over them. Far from corralling it, they're barely stirring it. As they pass, the oil—seemingly all of it—remains.

Louisiana lives by oil and by seafood. But oil rules. Fishing has nothing like the cash, the lobbyists, the destructive sophistication of the pusher to whose junk we're all addicted. But Florida lives largely by the whiteness of its sand. It has long eschewed oil. And the difference in what politicians will and won't say about oil is stark.

Florida's Governor Charlie Crist returns from a little airtime over the Gulf. His message: "It's the last thing in the world I would want to see happen in our beautiful state." He adds, "Until you actually see it, I don't know how you can comprehend and appreciate the sheer magnitude of that thing. It's frightening. . . . It's everywhere. It's absolutely unbelievable." Where oil money rules, governors are not at liberty to disclose such impressions. They're probably not at liberty even to think them.

"The president is frustrated with everything, the president is frustrated with everybody, in the sense that we still have an oil leak," says a White House spokesman.

But we've only just begun.

When Obama announced that he was opening up large new areas for offshore drilling, he said, "Oil rigs today generally don't cause spills. They are technologically very advanced." That's exactly right. Leaks, spills, and blowouts are never expected. Yet we know they happen. That should make us thorough in preparedness. But the human mind lets down its guard if big danger seems rare and remote.

And so we did. In 2009, the Interior Department exempted BP's Gulf of Mexico drilling operations from a detailed environmental impact analysis after three reviews concluded that a massive oil spill was "unlikely." Oil rig operators usually must submit a plan for how they'll cope with a blowout. But in 2008, the Bush administration relaxed the rules. In 2009, the Obama administration said BP didn't need to file a plan for how it would handle a blowout at the Deepwater Horizon. Now a BP spokesman insists, "We have a plan that has sufficient detail in it to deal with a blowout."

Obviously, they don't. Obviously, they aren't.

"I'm of the opinion that boosterism breeds complacency and complacency breeds disaster," says Congressman Edward Markey, a Democrat from Massachusetts. "That, in my opinion, is what happened."

Bush and Cheney's ties to big oil and their destruction of Interior oversight are infamous. But Obama's Interior secretary's ties to big energy also make environmentalists uneasy. As a senator from Colorado, Ken Salazar accepted some of BP's ubiquitous campaign contributions. In 2005, Senator Salazar voted against increasing fuel-efficiency standards for cars and trucks, and voted against an amendment to repeal tax breaks for ExxonMobil and other major petroleum companies. In 2006, he voted to expand Gulf of Mexico drilling. And then as Interior secretary, he pushed for more offshore drilling. The Interior secretary now says there have been well more than 30,000 wells drilled into the Gulf of Mexico, "and so this is a very, very rare event." The oil from those offshore rigs accounts for 30 percent of the nation's domestic oil production, he notes, adding, "And so for us to turn off those spigots would have a very, very huge impact on America's economy right now."

Probability, however, tells us that the spill's a very, very inevitable event. *Especially* with 30,000 wells drilled and roughly 4,000 wells currently producing oil in the region. In 2007, the federal Minerals Management Service examined 39 rig blowouts that occurred in the Gulf of Mexico between 1992 and 2006. So a blowout every four and a half months. I guess most are quickly controlled. Why aren't we ready for one that isn't? A car accident is a rare event, but we use our seat belts and we like to know we have air bags.

Rather than plan for the worst, Big Petroleum has indulged in—and been indulged by—a policy of waving away risks. In a 2009 exploration plan, BP strongly discounted the possibility of a catastrophic accident. A Shell analysis for drilling off Alaska asserts that a "large liquid hydrocarbon spill [hydrocarbon meaning oil and gas] . . . is regarded as too remote and speculative to be considered a reasonably foreseeable impacting event."

Foresee this: if you think it's difficult to clean up oil in the warm, calm Gulf of Mexico, imagine trying to do it in Arctic waters with icebergs, frozen seas, and twenty hours of darkness.

Speaking of the cold and the dark, "Drill, baby, drill" queen Sarah Palin just has to say *something* about all this. So she says we shouldn't trust "foreign" oil companies such as BP. She says, "Don't naively trust—verify." Verified: her husband worked for BP for eighteen years. Palin blames "extreme environmentalists" (c'mon, Sarah, is there any other kind?) for causing this blowout because they've lobbied hard to prevent new drilling in Alaska. If you follow what she has in place of logic, it could seem she'd rather that this had happened in her home state.

In and out of the comedy of horrors strides BP CEO and court jester Tony Hayward. "I think I have said all along that the company will be judged not on the basis of an accident that, you know, frankly was not our accident." That's what he actually says. Highlighting the failed blowout preventer, Hayward says, "That is a piece of equipment owned and operated by Transocean, maintained by Transocean; they are absolutely accountable for its safety and reliability."

Transocean's president and CEO says drilling projects "begin and end with the operator: in this case, BP."

Take that.

He says that Transocean finished drilling three days before the explosion. And he says there's "no reason to believe" that the blowout preventor's mechanics failed. That's what *he* actually says.

Halliburton's spokesman says his company followed BP's drilling plan, federal regulations, and standard industry practices.

In sum, BP has blamed drilling contractor Transocean, which owned the rig. Transocean says BP was responsible for the well's design and pretty much everything else, and that oil-field services contractor Halliburton was responsible for cementing the well shut. Halliburton says its workers were just following BP's orders, but that Transocean was responsible for maintaining the rig's blowout preventer. And the Baby Bear said, "Somebody's been sleeping in *my* bed."

By the end of the first week of May, a heavy smell of oil coming ashore along parts of Louisiana's coast begins prompting dozens of complaints about headaches, burning eyes, and nausea.

Meanwhile, I hear on the radio that in an effort to *do* something, "People from around the world have been giving the hair off their heads, the fur off their pets' backs, and the tights off their legs to make booms and mats to mop up the oily mess spewing out of the seabed of the Gulf of Mexico." Whether any of this stuff was ever actually used, I can't say. I never saw any; that I can say.

By May's second week, heavy machinery, civilian and military dump trucks, Army jeeps, front-end loaders, backhoes, and National Guard helicopters are pushing up and dropping down sand to keep an impending invasion of oil from reaching the marshes in and around Grand Isle, at the tip of Louisiana. Much of the mobilization falls to the Marine Spill Response Corporation, formed in 1990 after the *Exxon Valdez* disaster and maintained largely by fees from the biggest oil firms. Its vice president of marine spill response says that most of its equipment, including booms and skimmers, was bought in 1990. She says, "The technology hasn't changed that much since then."

"This is the largest, most comprehensive spill response mounted in the history of the United States and the oil and gas industry," crows BP's CEO Tony Hayward, sounding proud when he ought to be aghast and horrified by the scale of the mess and the upheaval.

Workers farther inland are diverting fresh water from the Mississippi River into the marshlands, hoping the added flow will help push back any oily water that comes knocking. "We're trying to save thousands of acres of marsh here, where the shrimp grow, where the fin-fish lay their eggs, where the crabs come in and out," says the director of the Greater Lafourche Port Commission. Enough Mississippi River water to fill the Empire State Building is now rushing into southeastern Louisiana wetlands *every half hour.* "We have opened every diversion structure we control on the state and parish level to try to limit the oil approaching our coasts," says the assistant director of the Office of Coastal Protection and Restoration, "nearly 165,000 gallons every second."

"It can't hurt," says a wetlands ecology professor at Ohio State University and an authority on the Mississippi's interaction with the Gulf of Mexico. Oh, but it can.

New fear factor: hurricane season. The image: hurricanes that could "churn up towering black waves and blast beaches and crowded cities with oil-soaked gusts." The news stories carry attributions such as "experts warned." And precise-sounding imprecision like: "As hurricane season officially starts Tuesday . . ." and "Last month, forecasters who issue a closely watched Colorado State University seasonal forecast said there was a 44 percent chance a hurricane would enter the Gulf of Mexico in the next few months, far greater than the 30 percent historic average."

To the ambiguity and imprecision, add unnecessary intonations of worrisome complexity: "The high winds may distribute oil over a wide area," says a National Hurricane Center meteorologist, adding, "It's a complex problem that really needs to be looked at in great detail to try to understand what the oceanic response is when you have an oil layer at the sea surface."

To most normal people faced with the real event of an out-of-control mess, especially people who've survived hurricanes, that kind of noninformation stokes anxiety, provokes fear—and gives no one a clear clue about what to do. Does one make decisions based on the difference between 30 and 44 percent? News you can use, it isn't. It's news that can help ruin your health.

Insult to injury: "Safety first," says a BP spokesman. "We build in hurricane preparedness, and that requires us to take the necessary precautions."

By mid-May, something like 10,000 people are Being Paid for cleanup efforts around the Gulf. BP has little choice. Anything less, there'd be riots. Most are fishermen riding around looking for oil, dragging booms that don't collect much oil, or putting out booms that can work only on oil that hasn't been dissolved by dispersants. About a million and a half feet—roughly 300 miles—of boom is already out along the coast. Other people are out picking oil off beaches with shovels.

How much oil are we dealing with? This gets good: Purdue professor Steve Wereley performs computer analyses on the video of the leaking oil to see how far and how fast particles are moving (a technique called particle image velocimetry). His conclusion: the well is leaking between 56,000 and 84,000 barrels daily. His other conclusion: "It's definitely not 5,000 barrels a day."

Just a few days ago, during congressional testimony, officials from BP, Transocean, and Halliburton estimated a "worst-case" scenario maximum flow of 60,000 barrels a day. Yet a BP spokesman says the company stands by its estimate of 5,000 barrels per day. There's "no way to calculate a definite amount," he says, adding coyly, "We are focused on stopping the leak and not measuring it."

That's Bull Poop. As the director of the Texas A&M University's geochemical and environmental research group points out: "If you don't know the flow, it is awfully hard to design the thing that is going to work."

Killing the well is proving difficult. Killing public confidence is easier. The fact that the real flow will turn out to be sixty times what BP was first saying, and twelve times the Coast Guard's most oft-repeated estimate, does the trick handily.

LATE MAY

Another discovery, another debate, more resistance from BP about disclosing how much oil is leaking. Scientists from the University of Georgia, Louisiana University, and elsewhere, aboard the research vessel *Pelican,* report finding—well, let them tell you: "There's a shocking amount of oil in the deep water, relative to what you see in the surface water," says Samantha Joye, a researcher at the University of Georgia. "There's a tremendous amount of oil in multiple layers, three or four or five layers deep in the water column." "Tremendous" meaning plumes as large as 10 miles long, 3 miles wide, and up to 300 feet thick. She reports methane concentrations up to 10,000 times higher than normal.

Dr. Joye says oxygen near some of the plumes has already dropped 30 percent because of oxygen-using microbes feeding on the hydrocarbons. In an e-mail, Joye calls her findings "the most bizarre-looking oxygen profiles I have ever seen anywhere." She notes, "If you keep those kinds of rates up, you could draw the oxygen down to very low levels that are dangerous to animals." Some parts of the plume had oxygen concentrations just above levels that make areas uninhabitable to fish, crabs, shrimp, and other marine creatures. "That is alarming," she says, adding that some of the Gulf's deepwater corals live directly below parts of the oil slick, "and they need oxygen."

But what are "plumes"? We don't quite get from the scientists an indication of how dense they are. Are we talking murky clouds? Tiny amounts detectable only with instruments? Enough to kill plankton? Small fish? Even a 10-by-3-mile plume is only a fraction of the Gulf. Are there other, more widespread plumes?

Plume definition: something flowing within a different medium.

The farther from the source of the blowout, the less concentrated the oil and gas hydrocarbons in the plume. Currents determine where plumes go. Dispersants also break up oil. If oil gets broken into very small droplets, say 1 micron in size, it no longer floats but dissolves into the seawater. "Clean" seawater contains oil concentrations less than 1 part per billion; in every billion drops of seawater there's less than one dissolved drop of oil. A polluted place, like a city harbor, may have between 100 and 800 drops of oil for every billion drops of seawater. You can also think of it as, say, 100 to 800 gallons of oil in every billion gallons of seawater. A billion gallons occupies about 5,000 cubic yards, or a space 100 yards long, 50 yards wide, and 1 yard deep. Two hundred gallons of oil would take up one cubic yard.

A bit of comparison. The Ixtoc blowout discharged a maximum of about 30,000 barrels per day. The present blowout is leaking perhaps twice that amount, so at any given distance from the blowout, our concentrations should be higher. Within a few hundred yards of the Ixtoc blowout, oil was 10,000 parts per billion. About 50 miles away, it was 5 parts per billion. Because the concentration gets lower and lower as oil travels away from the blowout, nobody knows much about how toxic the plumes of oil are.

They used to say "dilution is the solution to pollution," and sometimes that's right. If you think of a pollutant as anything that overwhelms the environment's ability to harmlessly absorb it, dilution can work. (Animals' bodies, though, often reconcentrate pollutants, much to their harm, so dilution isn't always the solution.)

BP's chief operating officer continues insisting that there exist no underwater oil plumes in "large concentrations." He says that this may depend on "how you define what a plume is here."

The way University of Georgia researcher Dr. Samantha Joye defines a plume, one large concentration of hydrocarbons from the blowout stretches at least 15 miles west of the gushing oil well, 3,600 feet beneath the sea surface, 3 miles wide, and up to 1,500 feet thick. The way University of South Florida researchers define a plume, there's an even larger plume stretching more than 20 miles northeast of the oil well, with the hydrocarbons separated into one layer 1,200 feet below the surface and another 3,000 feet deep.

"The oil is on the surface," says BP's somnambulant CEO Tony Hayward. "There aren't any plumes." That's certainly interesting, coming from someone whose company is blasting dispersants into the oil right at the seafloor as it emerges from the broken pipe, to keep the oil below the surface, and sending planes to carpet-bomb the slicks with more dispersant to sink the oil that has risen.

So how do *you* define plumes? In some places there's enough oil to discolor the water. But in most places, water samples come up clear. Yet the dissolved hydrocarbons show up vividly on instruments, and in some samples you can smell them. That's plumes.

Plumes are more evidence that there's more oil leaking than we're seeing. Because the Purdue University estimate and these reports of massive plumes seem so at odds with BP's continual 5,000-barrel-a-day drumroll, scientists want to send sophisticated instruments to the ocean floor to get a far more accurate picture of how much oil is really gushing from the well.

"The answer is no to that," a BP spokesman says. "We're not going to take any extra efforts now to calculate flow there."

That's just outrageous. Now I'm really angry. How does BP get to decide who can have access to the seafloor? The corporation has a permit to drill, and a responsibility to clean up its mess. Why does the government keep deferring to it? Why is it allowed to dictate what does and doesn't happen on public property? Why do our public agencies keep allowing it do whatever it wants?

There's at least one quite likely explanation for the discrepancy between what the Purdue scientist measured from the video of the leaking pipe and what BP is saying about what's on the surface.

And here it is: "It appears that the application of the subsea dispersant is actually working," says BP's chief operating officer. "The oil in the immediate vicinity of the well and the ships and rigs working in the area is diminished from previous observations."

And, considering that all the oil we can't see is polluting the Gulf, we can reasonably ask, In what way is that "working" for BP?

"The amount of oil being spilled will help determine BP's liability," confirms retiring U.S. Coast Guard admiral Thad Allen. But despite "overseeing the operation" on our behalf, Allen seems to be doing nothing—incredibly enough—to ensure that we actually send down some instruments designed to get the best possible estimate of *how much oil.*

I am not impressed with the Coast Guard so far. Admiral Thad Allen becomes to me a one-dimensional government talking head: the Thadmiral. Does he deserve to be a caricature? Of course not; does anyone? But in my anger, that's what happens.

Under the Clean Water Act, penalties are based on the number of barrels deemed spilled. Those penalties range from $1,100 to $4,300 a barrel, depending on the extent of the company's negligence. At, say, 5 million barrels, and if BP were found willfully negligent, it could face a fine of over $20 billion. So, yes, the dispersant is "working." Get it?

Dispersants begin accumulating well-deserved criticism. When broken up by dispersants, "The oil's not at the surface, so it doesn't look so bad," says Louisiana State University veterinary medicine professor Kevin Kleinow, "but you have a situation where it's more available to fish." By breaking oil into small particles, dispersants make it easier for fish and other sea life to soak up the oil's toxic chemicals. That can impair animals' immune systems, gills, and reproductive systems.

Marine toxicologist Dr. Susan Shaw says the dispersant "is increasing the hydrocarbon in the water." Dr. Samantha Joye says, "There's just as good a chance that this dispersant is killing off a critical portion of the microbial community as that it's stimulating the breakdown of oil." Louisiana State University environmental chemist Ed Overton is of the opinion that "we've gone past any normal use of dispersants." LSU's Robert Twilley wryly observes, "There are certain things with dispersants that are of benefit, and there are negatives, and we're having problems evaluating those trade-offs."

News flash! A new study shows that dispersant is no more toxic to aquatic life than oil alone. Okay, thanks, but that's not the question. The question is this: Is the mix more toxic to marine life than either alone? Part of the answer to that is: Yes, the mix is more toxic than dispersant alone, at least in lab tests. On the other hand, will dispersant, as its proponents insist, help speed the oil's degradation into harmlessness? Maybe; but will it also speed oil-caused mortality first? The problem is, no one really knows what will happen out in the complex Gulf. Everyone's guessing, and at best there are, indeed, trade-offs.

"This is what we call the Junk," Captain Keith Kennedy says derisively as he steers us from Venice, Louisiana, through the Industrial Canal. It's hard to imagine a more awful waterfront, and the whole place smells like petroleum. That's one reason a good chunk of southern Louisiana is called Cancer Alley.

The angled light, yellow through the heavy haze of moist air, joins the sounds of gulls and engines to make a Gulf morning. Once upon a time there was a wild coast here. Must have been magnificent.

Until a couple of weeks ago, Kennedy fished for redfish, "specs"—sea trout—and tarpon.

"When I first heard about the blowout, I didn't really think much about it. These things do happen. I figured they'd have it under control pretty quick. Their big metal thing didn't seem to work. I have to believe they're doing the best they can."

There is no single Mississippi River mouth. The mighty, muddy Mississippi speaks in tongues; her song is a chorale, her delta is a polyglot of channels. We're gonna run down via the channel called the Grand Pass, and from there through the Coast Guard Cut to East Bay, directly confronting the open Gulf of Mexico.

We pass a swimming alligator. A river otter pops its head up briefly.

A peregrine falcon comes high over the distant marsh, assessing the shorebirds for any weakness.

If oil comes into any of these channels, there is no way people can clean it from the intricate intimacy of these marshes.

One area with many resting birds is boomed. But birds fly. We're seeing pelicans, gulls, and terns diving. There's a slick near the diving birds. It could be natural. I don't see or smell any oil. There's no rainbow sheen or scent. It's very thin on the water, and at first Kennedy and I both think it's a slick from a school of fish below.

But then he spots some floating flecks that don't look familiar. Using a bait net, Captain Kennedy collects one, then a larger blob. The stuff's a bit gooey, the consistency of peanut butter but stickier. It smells like petroleum. It doesn't dissolve in water. It's hard to get off our hands. This is our first actual contact with the actual crude oil. It's nasty stuff. Let's hope we don't explode.

We pass a shack called Paradise, and another called Happy Ending. By now everything seems laden with portent; every sight and sign seems ominously like some metaphor of the all too real.

A rather gratifying amount of public and media interest arises over the fate of the magnificent bluefin tuna, which grows to over half a ton and whose numbers have been demolished by overfishing. Swimming at highway speeds, they tunnel throughout the whole Atlantic, but when spawning on our side of the ocean they migrate into the Gulf of Mexico. And this is spawning season. And though they can live for decades and grow to fifteen hundred pounds, they start out by hatching from millions of minute eggs to begin life as tiny drifting larvae. At the Gulf Coast Research Laboratory, a biologist opines, "This places the young larvae, I think, in a precarious position in respect to the location and magnitude of the spill." A tuna plankton expert here adds, "Large numbers of bluefin tuna larvae on the western edge of the Loop Current might be impacted by the oil spill as they move northward through the loop." Finding and counting fish larvae is painstaking work. They'll have to compare this year's numbers to prior and subsequent years. It'll take a while to learn more.

Understanding accrues slowly. Reactions happen at a different tempo. In the third week of May, the government suddenly closes 46,000 square miles to fishing, or about 19 percent of the Gulf of Mexico's federal waters.

The Louisiana Seafood Promotion and Marketing Board whistles in the dark that seafood from the areas not closed is still both available and safe; and more than half the state's oyster areas remain open. But as a local seafood market owner says, "Perception is everything."

And here's a different perception of the whole situation. The delta's Native Americans include the Pointe-au-Chien tribe. A century ago, Natives like them, isolated, illiterate, non-English-speaking, unable to get to New Orleans, missed the opportunity to claim land and territory after the Louisiana Purchase, in 1803—even though it had for millennia all been theirs. Much of southern Louisiana was claimed by the federal government, which auctioned a lot of it to land companies in the 1800s. Later, oil companies bought much of southern Louisiana. They swindled the Natives out of any crumbs they'd gotten.

Native Americans have in the past accused land grabbers and oil companies of seizing waterlands that rightfully belonged to them. They sued to regain vast tracts now owned by big landholding and energy companies. Needless to say, they lost.

"If you see pictures from the sky, how many haphazard cuts were made in the land, it blows your mind," says Patty Ferguson of the Pointe-au-Chien tribe. "We weren't just fishermen. We raised crops, we had wells. We can't anymore because of the saltwater intrusion." Sixty-year-old tribal elder Sydney Verdin feels a tingle of vengeful satisfaction. "I'm happy for the oil spill. Now the oil companies are paying for it the same way we've had to pay for it," says Verdin. "I can't think of one Indian who ever made any money from oil."

May 16. Out of the galaxy of goofy ideas, one that seems positively prosaic: they'll stick a tube into the leaking pipe. Why didn't any of us think of that? (Actually, of course, we did.) But their leaky tube is half-assed and it less-than-half works.

Incessant national airing of live video shows a lot of oil streaming past the mile-long tube sucking some of the oil from the ruptured pipe up to a waiting ship. BP is in an interesting bind. They say they're collecting 5,000 barrels a day through the tube. But for weeks they've been saying the well is leaking 5,000 barrels a day. Yet we can all see clearly on TV that most of the gushing oil isn't going into the tube. Busted!

A BP spokesman washes clean, sort of, fessing, "Now that we are collecting 5,000 barrels a day" through the tube, "it [the amount coming out of the pipe] might be a little more than that."

A *little* more? It's *twelve times* more.

Spin cycle: "From the beginning," intones the BP spokesman, "our experts have been saying there really is no reliable way to estimate the flow from the riser, so we have been implementing essentially a response plan." Anyone Buying Propaganda?

It gets better. Two days later, BP decides to announce that it is *not* collecting 5,000 barrels through the tube. "We never said it produced 5,000 barrels a day," says BP's chief operating officer. "I am sorry if you heard it that way." Oh, it's *our* mistake; we all heard it wrong. He says the tube is scarfing more like 2,000 barrels a day.

Lying? I don't know. But, well, actually, yes, since they *did* say, "we are collecting 5,000 barrels a day."

It would make BP look better if its spin doctors can convince us they're collecting less than 5,000 daily, whether it's true or not. Trouble is, we can't tell. Question is: Can we trust BP?

"We cannot trust BP," says Congressman Edward Markey. "It's clear they have been hiding the actual consequences of this spill." Purdue Professor Wereley, who'd estimated that the flow is more like 56,000 and 84,000 barrels daily, says, "I don't see any possibility, any scenario under which their number is accurate."

Ian MacDonald of Florida State University, an oceanographer who was among the first to question the official estimate, concludes that BP is obstructing an accurate calculation. "They want to hide the body," he says. Notes Congressman Henry Waxman, who chairs the

House Committee on Energy and Commerce: "It's an absurd position that BP has taken, that it's not important for them to know how much oil is gushing out."

During the third week of May, Tony Hayward says, "Everything we can see at the moment suggests that the overall environment impacts of this will be very, very modest."

When CNN's Candy Crowley asks Thad Allen for his response, the admiral responds, "Obviously they are not modest here in Louisiana. We don't want to perpetuate any kind of notion at all that this is anything less than potentially catastrophic for this country."

Crowley responds with what most of us are thinking: "Well, this is why people don't really trust BP, because here is the CEO of the company out there saying, 'We think the environment impact will be very modest.'"

Allen adds, "We're accountable. And we should be held accountable for this. We are taking this very, very seriously."

Thank you, Thadmiral. It's about time we got a government message in a plain-paper wrapper, and that we got to hear you come right out and say that.

In brown and orange globs, in sheets thick as latex paint, oil begins coating the reedy edges of Louisiana's wetlands. As crude oozes in the entrances, hope flees out the back door.

The worst fears of environmental disaster are, it seems, being realized. "Twenty-four miles of Plaquemines Parish is destroyed," rages a despairing Billy Nungesser, head of the parish. "Everything in it is dead. There is no life in that marsh. It's destroying our marsh, inch by inch." And because this is not a spill but an ongoing eruption, his prognosis: more of the same, coming ashore for weeks and months. Louisiana's governor says, "This is not sheen, this is heavy oil." He also fears that "this is just the beginning." He wields the statistics at stake: 60,000 jobs in Louisiana's $3 billion fishing industry; that Louisiana produces 70 percent of the Gulf's seafood, nearly one-third of the continental United

States' seafood. In addition, throughout the Gulf of Mexico region in a typical year, commercial fishermen usually catch more than 1 billion pounds of fish and shellfish, and nearly 6 million recreational fishermen make 25 million fishing trips. (One cannot help wondering whether the sea creatures would rather face our oil or our nets and hooks.)

The EPA tells BP: you have a twenty-four-hour deadline to choose a less toxic chemical dispersant. Dispersants have gotten our attention, but plenty of other chemicals—many of them similar—drain from America's Heartland to America's Wetland and beyond. Soaps, dishwashing liquids, and industrial solvents are all oil dispersants. Down the drain they go. Household cleaners, ingredients used to make plastics, and pesticides. Medical residue that goes from body to potty to Gulf. Livestock waste and traces of the drugs they've been given. Enough estrogen from birth control pills to bend genders and mess the sex of fish. Caffeine. Herbicides toxic to aquatic animals, by the thousands of tons. And good old (actually new, synthetic) fertilizers that cause algae populations to skyrocket, leading to an explosion of the bacteria that decompose them, which depletes the deeper water's oxygen, killing everything. That's the "dead zone" that happens every year in the Gulf (and now in hundreds of other coastal places around the world and is coming soon to a river mouth near you). The Gulf dead zone holds the distinction of first and worst, and this year it's set to break a record: it's about 8,000 square miles, the size of New Jersey. Or Massachusetts. (Some of us remember when it was a cute little dead zone no bigger than Delaware.) There's actually a federal goal of shrinking it to less than 1,900 square miles by 2015. Agencies planned to accomplish that by getting midwestern farmers to put less fertilizer into the river system. But good luck. "It's getting bigger over the years, and it's extending more into Texas," says Nancy Rabalais, director of the Louisiana Universities Marine Consortium. So one more thing is killing the Gulf, and it's a big one: agriculture. Modern, industrialized, artificialized, corporatized, heavily lobbied agriculture. When the oil is gone, the water of the mighty Mississippi will remain all too fertile for the Gulf's good.

For vigilance, we the people pay taxes that support our government's defense of our interests. Among the vigilant defenders is that government agency called the Minerals Management Service, a branch of the Interior Department. Its mission, if it decides to accept it: make sure extraction of oil and other nonliving resources is done well, done safely, and done to certain specified standards. And yet in at least one region—though there's no evidence that this was an issue in the Gulf—the MMS got a little too informal when it mattered. In 2008, it came to light that eight MMS employees had accepted lavish gifts and had partied with—and in some cases had sex with—employees from the energy companies they regulated. A formal investigation found the agency's Denver office rife with "a culture of substance abuse and promiscuity." The report further noted, "Sexual relationships with prohibited sources cannot, by definition, be arms-length."

On May 19, Interior Secretary Salazar announces that he's dividing the disgraced Minerals Management Service into three units. One might say he's dispersing the agency. The agency had been created by Ronald Reagan's infamous Interior secretary James Watt. That explains some things. The current agency's three missions—energy development, enforcement, and revenue collection—"are conflicting missions and must be separated," Salazar says. Applause. No more sex, drugs, and rock-and-oil. In a few days the agency's chief will quit under pressure.

The agency has been collecting royalties from the companies it regulates. That lowers the incentive for strict safety oversight. The new idea: there'll be a Bureau of Safety and Environmental Enforcement (this one will inspect oil rigs and enforce regulations), a Bureau of Ocean Energy Management, Regulation and Enforcement (this one will oversee offshore drilling leasing and development), and an Office of Natural Resources Revenue (to collect the billions of dollars in royalties from mining and drilling companies extracting resources from American territory).

In Grand Isle and Barataria Bay, Louisiana, hideously oiled gulls and pelicans struggle to keep the life they'll lose. Parent birds have brought to their eggs coatings of oil, blocking oxygen from entering the shell. Chicks that manage to hatch will know only a short life on the gummy surface of a petroleum-coated planet. Behind the lines of useless boom, oil coats the marsh cane.

"It took hundreds of years to create this," one fisherman says, low-balling the time required by about six thousand years. "And it's gone just like that."

Meanwhile, President Obama announces that automakers must meet a minimum fuel-efficiency standard of 35.5 miles a gallon by 2016. Savings over the five-year phase-in: 1.8 billion barrels of oil. That means saving a million barrels of oil daily, about seventeen times faster than it's leaking from the blowout.

Gulf breezes smell of oil. Marshes smell of oil. And "All systems are go" for the seventy-fifth annual Louisiana Shrimp and Petroleum Festival. Says the festival's director, "We will honor the two industries as we always do."

On May 24 in Port Fourchon, Louisiana, seven Greenpeace members board the ship *Harvey Explorer* that's heading north in July to support drilling operations in the Arctic. In oil from the blowout, they write, "Is the Arctic Next?" on the hull. They're all charged with *felonies.*

No one from BP, Transocean, or anyone else has been charged with anything. That's our government Bullying People. (It'll take about three months for the squeaky wheels of justice to dismiss the charges.)

A house divided: Coast Guard admiral Thad Allen says on May 23 that BP's access to the mile-deep well means the government could not take the lead to stop the leak. Yet a few hours later outside BP's Houston headquarters, a tough-talking Interior secretary Salazar (who early in the crisis vowed to "keep the boot on the neck" of BP) says, "If we find they're not doing what they're supposed to be doing, we'll push them out of the way." This prompts the Thadmiral to wonder out loud to reporters, "To push BP out of the way would raise the question of, 'replace them with what?'" He adds, "They have the eyes and ears that are down there."

Okay, at least now he's pretty much acknowledged who's really in charge. Fact is, the government, as the *Christian Science Monitor* puts it, is "incapable of taking over from BP at the wellhead and unwilling to displace the web of contractors leading the cleanup at BP's behest."

Allen remarks about BP, "They are necessarily the modality by which this is going to get solved."

In my experience, people who use the word "modality" never quite get to the point.

He gets to the point: "They're exhausting every technical means possible to deal with that leak. I am satisfied with the coordination that's going on."

I'm not. Nor is anyone else I know. A few days later, even the president will say he regrets not realizing that oil companies did not "have their act together when it came to worst-case scenarios." He will add a stern admonishment: "Make no mistake, BP is operating at our direction."

It's a mistake I will continue to make—often—in the upcoming weeks. Meanwhile, enraged over BP's stonewalling and its refusal to entertain new ideas and alternate solutions—or to make any seemingly sincere attempt to collect oil at the surface—Plaquemines Parish president Billy Nungesser rails, "BP has taken over the Gulf of Mexico, and we're doing nothing to stop them."

Indeed, it feels to many as if the Coast Guard has handed BP the keys to our car and climbed into the back seat.

Louisiana's governor declares a commercial fisheries' failure to trigger aid. Within a week, the fisheries disaster declarations spread to include Alabama and Mississippi. Aid means tax dollars. Aid means that oil is not as cheap as it seems at the pump. We pay anyway.

Here's who doesn't get dispersants: the head of the EPA. Lisa Jackson, Environmental Protection Agency chief, exclaims, "Oh my God, it's so thick!" as she assesses a cupful of the oily mess dipped from the mouth of the Mississippi. "At a minimum what we can say is dispersants didn't work here," Jackson actually says. She adds, "When you see stuff like this, it's clear it isn't a panacea."

Panacea? As if perfection would be to just send all the oil out of sight, out of mind? Isn't that BP's dream scenario, to make it all seem to just go away by sinking it all below the surface? I don't agree. At a minimum, what we *can* say is that dispersants don't work if your goal is to avoid polluting the water on a massive scale.

Panacea? Dr. Susan Shaw of the Maine-based Marine Environmental Research Institute says, "The worst of these dispersants—sold by the name Corexit 9527—is the one they've been using most. It ruptures red blood cells and causes fish to bleed. With 800,000 gallons of this, we can only imagine the death that will be caused." But that's the problem: we can only imagine. It causes harm in laboratory tests at certain concentrations, but the dose makes the poison, and we have no clarity on what it's doing in the Gulf.

Panacea? The *Exxon Valdez* disaster is what first linked Corexit to respiratory, nerve, liver, kidney, and blood disorders. *Exxon Valdez* cleanup workers reported blood in their urine. EPA data shows Corexit more toxic and less effective than other approved dispersants.

A BP spokesman calls Corexit "pretty effective," adding, "I'm not sure about the others." BP's main reason for continuing to use Corexit appears to be its close ties to the manufacturer.

Panacea? Dr. Shaw writes in the *New York Times* after actually diving in part of the dispersant-and-oil mixture, "What I witnessed was a surreal, sickening scene beyond anything I could have imagined." She describes the murky mixture drifting a few meters down, then concludes, "The dispersants have made for cleaner beaches. But they're not worth the destruction they cause at sea, far out of sight. It would be better to halt their use and just siphon and skim as much of the oil off the surface as we can. The Deepwater Horizon spill has done enough damage, without our adding to it."

The pressure seems to be pushing the EPA. Its officials finally obtain and make public a list of the concoctions' ingredients. One version of Corexit ("corrects it"; get it?) contains benzene and 2-butoxyethanol, linked to destruction of red blood cells and cancers in lab studies of monkeys, rats, mice, rabbits, and dogs, which can lead to kidney, spleen, or liver damage. Also it caused breathing difficulties, skin irritation, physical weakness and unsteadiness, sluggishness, convulsions, birth defects, and fewer offspring in mammals. This stuff, you *don't* want to swallow. The head of the Louisiana Shrimp Association calls Corexit "the Gulf's Agent Orange."

Now EPA administrator Lisa Jackson orders BP to take "immediate steps to scale back the use of dispersants" by 50 to 75 percent. (Well, which one is it?) While the government had approved the use of dispersants before this blowout, no one had anticipated that they'd ever be used at this scale and in these quantities.

My question is: Why are they using dispersants at all? As the incessant video shows, the entire leak is erupting from one small pipe. It's not like a massive tanker spill, where it's all in the water already and there's no ongoing "source." This is very different. They have their hands around this whole thing at that pipe. It seems to me it could *all* be captured. After all—*it's an oil well.*

And now: voider of her own election, former Alaska governor turned national misfortune Sarah Palin comes out of her nutshell again to say that—to make a long story short—Obama's response has been slow. It's another signature blast of her sound-and-fury insight. And it prompts White House spokesman Robert Gibbs to suggest that Palin needs her own personal blowout preventer.

Granted, the administration's response does seem slow. And the Coast Guard, in my estimation, has been disappointing. Our president disagrees with the likes of Palin and me, saying, "Those who think we were either slow on the response or lacked urgency, don't know the facts."

In a late-May press release, a group called Public Employees for Environmental Responsibility alerts us to the details and particulars of the BP response plan. Basically, there aren't any. Plus, it's so full of nonsense that apparently no regulator read it seriously.

Dated June 30, 2009, the "BP Regional Oil Spill Response Plan—Gulf of Mexico" covers all of the company's various operations in the Gulf. The plan lists "Sea Lions, Seals, Sea Otters, and Walruses" as "Sensitive Biological Resources" in the Gulf. None of those animals live there (at least not since the Caribbean monk seal went extinct in the 1950s). BP has obviously just cut-and-pasted from documents written for drilling in Alaska. The document also gives a Japanese home-shopping website as a "primary equipment providers for BP in the Gulf of Mexico Region for rapid deployment of spill response resources on a 24 hour, 7 days a week basis." The sea turtle expert you're supposed to call is a researcher who's been dead for five years. And the 600-page plan *never discusses how to stop a deepwater blowout.*

By May's last waning days, some fishermen hired to do cleanup by BP say they have become ill after working long hours near oil and dispersant. Headaches, dizziness, nausea. Difficulty breathing. Burning eyes. The EPA's air monitoring has detected odors strong enough to cause sickness. The EPA's website warns coastal residents that these chemicals "may cause short-lived headache, eye, nose and throat irritation, or nausea."

BP says it's unaware of any health complaints.

That's because: "You don't bite the hand that feeds you," says the president of the Commercial Fishermen's Association. Many fishermen have told *him* about feeling ill. "You left in the morning, you were OK. Out on the water, you've got a pounding headache, throwing up." And yet, he says, "BP has the opinion that they are not getting sick."

Maybe time for a second opinion.

On May 26 BP subjects us to two new vocabulary terms. Of all the kooky names for dopey ideas, these two take the cake: "junk shot" and "top kill." Let's use them in one sentence: "They don't work either." The rodeo names reflect the rodeo thinking that got us here. One half-baked idea after another.

The Bright Ploy this time: attempt to stop the upward flow of oil by sending heavy drilling fluid down the well. Workers have triggered the original blowout and explosion by *removing* the heavy drilling fluid that had, in fact, been holding down the oil. So we might call this the "oops" strategy, as in "Oops, let's go back to what was working before we caused the blowout."

It's *way* too late now, though. As the University of Texas's Petroleum Engineering Department chairman says, "You have the equivalent of six fire hoses blasting oil and gas upward and two fire hoses blasting mud down. They are at a disadvantage."

BP pegs its chance of success at 60 to 70 percent. (New definition of "pegged": made up, fabricated.) "We're doing everything we can to bring it to closure, and actually we're executing this top kill job as efficiently and effectively as we can," says BP chief operating officer Doug Suttles.

I don't know what to make of the word "actually" there. Is he surprised? Or does he know that we know that he's just BS-ing us?

The University of Texas's Petroleum Engineering Department chairman watches a live video and says, "It's not going well."

"I wouldn't say it's failed yet," says BP's chief operating officer. With dramatic flair that seems oblivious to the sheer irresponsibility it implies, BP reminds us that the method *has never been tried before at such depth.* "This is the first time the industry has had to confront this issue in this water depth," gushes BP's CEO, the ever-perspicacious Tony Hayward, "and there is a lot of real-time learning going on."

"We've never tried this before in water this deep" is a bad answer when you've been *drilling in water this deep.*

"It's a wait-and-see game here right now; so far nothing unfavorable. . . . The absence of any news is good news," says the chipper Thadmiral.

Thad Allen was a hero in the Hurricane Katrina disaster, yet I find myself unable to believe what he's telling us. Not entirely fair, but my anger is stoking my cynicism. The impression given—at least the impression I form—is that Admiral Allen is up to his neck in oil and over his head in this debacle. In my mind he becomes government chief of useless statements.

Next, Thad Allen begins talking enthusiastically about BP's planned "junk shot." "They're actually going to take a bunch of debris, shredded-up tires, golf balls, and things like that and under very high pressure shoot it into the preventer itself and see if they can clog it up and stop the leak."

There's that word "actually" again.

It all fails.

Obviously, Admiral Allen has been bugging me. But to be honest, I'm not sure how much of that is him—and how much is me. It's difficult to maintain one's objectivity amid so much subjective confusion. The drilling rig wasn't his responsibility, and he certainly didn't cause the blowout.

BP announces that it has spent $930 million responding to the spill. The company is acting like an emotionally distant husband seeking appreciation from his wife and children by telling them how much his bills cost.

And rather belatedly, Tony Hayward calls the blowout "a very significant environmental crisis" and a "catastrophe." After he'd earlier said the oil leaked has been "tiny" compared with the "very big ocean," his media coaches are earning their fees. But he should try telling us something we *don't* already know.

So far, about 26,000 people have filed damage claims.

Researchers on the University of South Florida College of Marine Science vessel *Weatherbird II* report discovering a massive amount of oil-polluted water beneath the Gulf of Mexico, in a layer hundreds of feet thick, down to a depth of well over 3,000 feet, drifting in a several-miles-wide plume stretching twenty-two miles from the leaking wellhead northeast toward Alabama. Chemical oceanographer David Hollander makes this announcement, saying with all due scientific caution that it's likely oil from the blown-out well.

But really, what else could it be? Let's review: Oil is gushing from a well. One end of the plume is in the vicinity of that well. And the company running the drilling operation has been spraying dispersants on the seafloor and at the surface to keep as much of the oil underwater as is humanly possible. Scientists have detected oil underwater, ergo—what else could it be from?

"This is when all the animals are reproducing and hatching, so the damage at this depth will be much worse," says Dr. Larry McKinney of the Harte Research Institute for Gulf of Mexico Studies. "We're not talking about adults on the surface; it will impact on the young—and potentially a generational life cycle. At the depth that these plumes are at, the sea will be toxic for God knows how long."

The federal government closes more fishing areas, to the west and south, on May 25. The closed area now: 54,096 square miles, over 22 percent of the Gulf of Mexico's federal waters. "This leaves approximately more than 77 percent still open for fishing," our National Marine Fisheries Service adds with silver-lining turn-a-frown-upside-down think-positiveness.

Meanwhile, U.S. Geological Survey director Marcia McNutt says the oil leaked in the last five weeks totals somewhere between 18 million and 39 million gallons. That's way past *Exxon Valdez*'s 11 million gallons. (So it's said; others insist the total was much more than Exxon ever admitted.)

And speaking of Alaska, Shell Oil has been poised to start exploratory drilling this summer as far as 140 miles off Alaska's coast. But now the Obama administration suspends proposed exploratory drilling in the Arctic Ocean until 2011. *Alaska politicians are pissed!* They have to be—about 90 percent of Alaska's general revenue comes from the petroleum industry. It's what helps get them elected. They're like sled dogs who start the day with a big bowl of oil and then get harnessed up to pull Petroleum's sled.

❧

"*I was* certain I was going to die," Deepwater Horizon survivor Stephen Stone tells a congressional hearing panel. He says the April 20 blast was "hardly the first thing to go wrong." He testifies, "This event was set in motion years ago by these companies needlessly rushing to make money faster, while cutting corners to save money." More than a day after the explosion, Stone was finally back on land. "Before we were allowed to leave, we were lined up and made to take a drug test. It was only then, 28 hours after the explosion, that I was given access to a phone, and was allowed to call my wife and tell her I was OK."

Then, a few days later, a representative of rig owner Transocean asked him to sign a document "stating I was not injured, in order to get $5,000 for the loss of my personal possessions." He declined to sign.

These are the kinds of people we're dealing with.

Eight workers airlifted to a Louisiana hospital this week were released. That's the good news. One fisherman hospitalized after becoming ill while cleaning up oil—severe headaches, nosebleeds, and so on—files a temporary restraining order in federal court against BP. He wants BP to give workers masks and not harass workers for publicly voicing their health concerns. He also says, "There were tents set up outside the hospital, where I was stripped of my clothing, washed with water and [had] several showers, before I was allowed into the hospital. When I asked for my clothing, I was told that BP had confiscated all of my clothing and it would not be returned."

A lot of fishermen are reluctant to complain. Making as much as $3,000 a day cleaning up the oil, they fear losing their jobs with BP. If it's partly hush money, BP's plan for them is working.

Of course, a BP spokesman says there have been no threats against workers for speaking out. He adds, "If they have any concerns, they should raise them with their supervisors."

In the space after that statement, I hear "and not with anyone else." When I ask one worker a question, he says to me, "No comment." When I ask him if his supervisors have told him to say that, he says, "Yeah."

"The only work fishermen can get right now is with BP," affirms the fisherman seeking the restraining order.

A fisherman's wife says her husband called her from a boat, saying, "This one's hanging over the boat throwing up. This one says he's dizzy, and he's feeling faint." She says they were downwind of it and the smell was "so strong they could almost taste it."

In addition to concern over oil, many fear the *million* gallons of dispersant served so far. The dispersant's own manufacturer states that people should "avoid breathing vapor" and that when this product is present in certain concentrations in the air, workers should wear masks.

The fisherman's wife says her husband came home so sick he collapsed into a recliner without eating dinner or saying hello to her or their children. After three weeks of coughing and feeling weak, he agreed to go for medical help. His wife's been trying to get BP to give the workers masks.

BP says workers who want to wear masks are "free to do so"—as long as they receive instructions from their supervisors on "how to use them."

A spokesman for the shrimpers' association insists that BP has told workers they are not allowed to wear masks: "Some of our men asked, and they were told they'd be fired if they wore masks." Environmental groups offering free masks to workers have been told by BP that they can't do that. If you wear a respirator you have bought with your own money, if you wear a respirator someone has given you—you're fired.

And here's why: oil is not their problem. Their problem is that they are eating. At least, that's what BP says. BP's CEO and chief harlequin Tony Hayward actually says, "Food poisoning is clearly a big issue." He adds, "It's something we've got to be very mindful of. It's one of the big issues." He himself seems to be suffering from foot-in-mouth disease.

"Headaches, shortness of breath, nosebleeds—there's nothing there that suggests foodborne illness," said Dr. Michael Osterholm of the University of Minnesota School of Public Health. "I don't know what these people have, but it sounds more like a respiratory illness."

The fisherman's wife has better data. She says there's no way her husband and the other men had fallen victim to food poisoning—they were on eight different boats and didn't eat the same food.

The director of Louisiana's Department of Health and Hospitals says, "It's hard to understand if nausea or dizziness or headache is related to the oil or to working in 100-degree heat." (I wonder why BP didn't think of that.) Wearing respirators could help unpack that. No, never mind; we're told that respirators could add to heat stress.

There's another reason the suffering workers aren't using respirators. But get ready for some tortured logic: the head of the federal Occupational Safety and Health Administration (OSHA) says the toxins in the Gulf air aren't concentrated enough to require workers to wear respirators. Based on that, BP says there are no health threats to workers. After public-health advocates criticize both of them, OSHA's head tells C-SPAN that he wouldn't advise using his agency's "out-of-date" guidelines. Confused? It can take your breath away.

Who else can't breathe? A Dauphin Island Sea Lab study finds a dramatic decline in dissolved oxygen near the ocean bottom at sites twelve and twenty-five miles off Alabama. The study's senior scientist, Dr. Monty Graham, announces, "Oxygen is dropping out offshore. We got minimum dissolved oxygen values of 1.7 micrograms per liter." Dissolved oxygen levels below 2 micrograms per liter are considered too low for almost everything that depends on oxygen for normal living. The values found are less than a fifth of normal. Dauphin Island Sea Lab director Dr. George Crozier says, "This is the kind of unexpected consequence that I warned BP representatives of on May 3rd, after they announced the successful application of dispersant at 5,000 feet."

And yet—. Other scientists will find the oxygen depletion rather moderate. And a federal panel of about fifty experts recommends continued use of chemical dispersants, saying populations of the underwater animals likely to be killed have a better chance of rebounding quickly than birds and mammals on the shoreline.

That's probably true, if you decide to be unconcerned about turtles, whales, and dolphins, and if you write off the possibility of effectively capturing floating oil. And at the heart of the matter are two things: dispersants are easy; dispersants make things look better. Plus, no one really knows for certain what would happen here with or without dispersants. There's a lot of guessing on the details.

On the final day of May, engineers begin trying to fit a new "top hat." The dome weighed 100 tons; this cap weighs two tons (further evidence that their shrinking thinking is all over the map). The idea now is that a diamond-bladed pipe cutter will create a smooth cut of the mangled pipe, right at the blowout preventer, to facilitate a tight fit.

Drawback 1: kinks in the pipe have been slowing the rate of leak; cutting the pipe will facilitate a higher flow. Before the pipe is cut, the government has estimated that the oil is flowing at 12,000 to 19,000 barrels a day—meaning at least 20 million gallons since April 21. Now BP says that cutting the kinked pipe will increase the flow by as much as 20 percent. (It'll be more like 300 percent.)

Coast Guard admiral Thad Allen says, "If we don't get as clean a cut as we want, then we'll put something called a 'top hat' over it, which is a little wider fitting, but you have an increased chance that some oil will come out around the sides." Definition of "increased chance": 100 percent certainty. And note: "we," thrice. Quite the sense of camaraderie, the admiral and BP. Boot on their neck or feather boa?

And that brings us to Drawback 2: the diamond blade gets stuck, and a pair of remotely operated hydraulic shears finish a rougher-than-intended job.

"It is an engineer's nightmare," says a Louisiana State University professor. "They're trying to fit a 21-inch cap over a 20-inch pipe a mile away using little robots. That's just horrendously hard to do." Tulane Energy Institute's associate director says it's like trying to place a tiny cap on an open fire hydrant.

BP is doing all it can. "No one wants this over more than I do," says BP CEO Tony Hayward. "I'd like my life back."

So would the families of the people who died on the rig. So would everyone in the Gulf region. Hello.

Meanwhile, an oil spill in Alaska has caused a shutdown of the Trans-Alaska Pipeline. And in a piece headlined "That Was Then, This Is Then," MSNBC's *Rachel Maddow Show* points out that in June 1979, an Alaska spill shut the pipeline at the same time oil gushed for months from a blown-out Mexican well in the Gulf of Mexico. "If you close your eyes and listen to the news reports from back then," she says, "you'd be forgiven for thinking you were listening to today's news." Back then, the well was being drilled by the same company that later changed its name to Transocean. Its blowout preventer had failed. The responses included spreading chemical dispersants, putting out booms along shorelines, and burning floating oil. People worried about "underwater plumes" and the possibility that currents would carry oil to Florida. Being Mexican, they did not have a "top hat"; they tried a "sombrero." When that failed, they tried forcing stuff down the top and pumping metal balls into the well to jam it up. And when that failed, a pair of *relief wells* finally stopped the blowout. Maddow's summation: "The stuff that did not work then is the same stuff that does not work now: same busted blowout preventer, same ineffective booms, same toxic dispersant, same failed containment domes, same junk shot, same top kill; it's all the same." Point being: nothing's changed in preparedness or response. What's changed is the depth. Ixtoc was in about 160 feet of water; the present blowing well starts a mile down. "All they've gotten better at," Maddow notes, "is making the risks worse."

Congress has certainly played a role—and been played—in encouraging risky drilling. One trend has been to undo the tapestry of prior protections. In 1995, Congress "decided" to reduce the government's royalties on oil and gas extracted from deep water (you can bet the idea came from Big Oil in an envelope marked "Campaign

Contribution"). The goal and result: more drilling encouraged. Okay, fine; bringing more of our oil under domestic production *is* good for national security. But other nations charge more. Oil taxes and royalties in the United States are considered much lower than elsewhere in the world. Our country should benefit financially, as the oil companies do. And what else isn't fine is that this encouraged more and riskier drilling but not more safety. Industry assurances that deepwater drilling was "safe" rocked a willing Congress to sleep on the issue. The fact that accidents are rare and unpredictable has substituted for the obvious truth and certainty that accidents do happen.

In 2001 the president's National Energy Policy report (it was actually the vice president's; the first page is Dick Cheney's submittal letter to George W. Bush) ordered agencies to increase oil production and remove "excessive regulations." The report has a lot of good ideas. For instance, it says, "A primary goal of the National Energy Policy is to add supply from diverse sources. This means domestic oil, gas, and coal. It also means hydropower and nuclear power. And it means making greater use of non-hydro renewable sources now available." But in practice, federal agencies didn't get past the first sentence on that list. Maybe they never really intended to; I don't know.

In 2005, despite high oil prices and even President George W. Bush saying oil companies needed no further drilling incentives, the Republican-dominated Congress again lowered the royalties oil companies are required to pay our national Treasury. That's nonsensical, especially in an era of massive federal deficits, but part of the ideology appears to be a desire to starve the government so there is scant money for wasteful social programs like education, health, and environmental protection. And while, yes, there are excessive regulations, there is also excessive greed. Regulations don't threaten business; they threaten greed, the greed that threatens both us and our nation's economy. In 2006, Louisiana's congressional delegation supported giving a share of oil royalties to states that allowed drilling. This means using national oil revenue directly to achieve state policies that benefit Big Oil. Nice giveback. Clever.

And while calling for more incentives to drill, baby, Congress slashed those annoying safety regs. From 2002 to 2008, Congress approved budgets reducing regulatory staff by over 15 percent. So we got more complex, deeper drilling, higher-volume oil pools, no further safety. In 2000, the Interior Department had voiced concerns that industry's extensive use of contractors and inexperienced offshore workers in deep water created new risks. *And* a 2004 Coast Guard study—repeatedly cited by Congress's own Congressional Research Service—warned, "Oil spill response personnel did not appear to have even a basic knowledge of the equipment required to support salvage or spill clean-up operations." Lawmakers slept peacefully. Environmental groups focused on maintaining the existing moratorium on new drilling, not operational safety or response. Consequently, regulatory proposals often drew fewer than ten "public" comments, but most came from the oil industry.

With democracy working on half of its cylinders, the Interior Department politely filled the public-interest vacuum with a new tendency to better serve, rather than better monitor, the oil companies. The person in charge of offshore drilling for Interior boasted that he "oversaw a 50 percent rise in oil production." But he was supposed to be a regulator, not a fixer. Ergo: deep blowout preparedness = zero.

EARLY JUNE

A dead dolphin rots in the shore weeds; an oil-stained gull stands atop its corpse. "When we found this dolphin, it was the saddest darn thing to look at," says the cleanup worker who is taking a news team on a surreptitious tour. "There is a lot of cover-up for BP. They specifically informed us that they don't want these pictures of the dead animals." The shore is littered with oiled marine creatures, some dead, others struggling. "They keep trying to clean themselves," the guide says. "They try and they try, but they can't do it. Some of the things I've seen would make you sick." He mentions that he recently found five turtles in oil. Three were dead. Two were dying. He says, "Nature is cruel, but what's happening here is crueler. No living creature should endure that kind of suffering."

News crews are now being barred from and escorted away from public beaches and public roads. The cops acknowledge that they're taking orders from BP. No one can figure out how or why this is being allowed, but as far as anyone can tell, it seems the Coast Guard is abetting BP.

On the first day of June 2010, BP is trying to cover the leak with the new top hat. Akin to applying condoms after they're pregnant. To defeat ice formation this time, technicians will inject heated water and methanol into the cap. To capture more oil, they begin closing the vents.

The Coast Guard Thadmiral tells us the goal is to gradually capture more of the oil. For anybody who didn't grasp that, he compares the process to stopping the flow of water from a garden hose with your finger: "You don't want to put your finger down too quickly, or let it off too quickly." A rather odd analogy. What does he mean?

After robots place the cap, video shows *plenty* of escaping oil still billowing around the cap's lip. "A positive step but not a solution," says the Thadmiral. "Even if successful, this is only a temporary and partial fix and we must continue our aggressive response operations at the—"

Okay, never mind. The plan is to capture most of the spewing oil and bring it up to a surface ship.

The Thadmiral says the "ultimate solution"—relief wells—is not likely till August.

Relief wells, alternate take: "The probability of them hitting it on the very first shot is virtually nil," says the president of the American Association of Petroleum Geologists, David Rensink, who spent thirty-nine years in the oil industry, mostly in offshore exploration. "If they get it on the first three or four shots they'd be very lucky."

BP shares lose 15 percent of their value on news that its attempted stop-from-the-top hasn't worked, indicating that the leak—and BP's liabilities for economic and environmental damages—will likely continue mounting for months.

The Justice Department announces criminal and civil investigations into the Gulf oil disaster. "All possible violations of the law," including the Clean Water Act, Oil Pollution Act, Endangered Species Act.

About 15,000 barrels of oil a day begin finding their way out the high end of the pipe and into the ship *Discoverer Enterprise.*

BP's Tony Hayward, sounding like he's trying to convince even himself, says the cap will likely capture "the majority, probably the vast majority" of the gushing oil. Ever the cheerleader for the sheer magnificence of the enterprise, he himself gushes, "It has been difficult to predict because all of this is a first. Every piece of this implementation is the first time it's been done in 5,000 feet of water, a mile beneath the sea surface."

Yes, Tony, that's what people mean when they say "total lack of preparedness."

As for BP's statement that the present cap might be capturing "the vast majority" of the spew, Purdue's Professor Wereley—who'd initially busted BP and the Coast Guard with his estimate that the blowout was spewing at least 56,000 barrels daily—says, "I don't see that as being a credible claim. I would say to BP, show the American public the before and after shots of the evidence on which they're basing that claim."

Similarly, a University of California researcher says, "I do not know how BP can make that assertion when they don't know how much oil is escaping." He believes that cutting the kinked riser pipe in order to install the cap increased the flow by far more than the 20 percent BP and government officials had predicted. He speculates it may be spewing what BP had called the "worst case": a 100,000-barrels-a-day blowout. He says the video feed now appears to show "a freely flowing pipe," adding, "From what it looks like right now, it suggests to me they're capturing a negligible fraction."

BP says "the vast majority," the academic says "a negligible fraction." Let's see if the Thadmiral can mediate. Ready? Go: "They continue to optimize production," Allen tells reporters. Then, digging himself an alternate escape route, he adds, "I have never said this is going well. We're throwing everything at it that we've got."

BP intends to hook up a new, tighter cap, with more pipes that will attach to more vessels. Its Bigger Plans include four collection vessels passing oil to two tanker ships. The collection vessels can process a combined total of between 60,000 and 80,000 barrels per day.

The discerning reader may note this subtlety: there's a slight difference among that planned capacity, the 1,000 barrels per day BP had first announced, its grudging acknowledgment and later insistence that 5,000 barrels were leaking, and its refusal to engage on the question of "how much" after Purdue professor Steve Wereley estimated 56,000 to 84,000 barrels. Yet BP plans to collect essentially the very same amount the professor estimated. What does that tell you?

Meanwhile, the Thadmiral says that within the next week all these activities "could take leakage almost down to zero." Says he's ordered a special task group to work up new estimates on how much oil is still gushing out. (It's about time.) Says, "I'm not going to declare victory on anything until I have the numbers." He adds for emphasis, "Show me the numbers."

Here's the only number we need: zero. Show us zero.

On June 9, BP shares hemorrhage an incredible 16 percent, to $29.20. BP shares have lost half their value, wiping off $90 billion in market capitalization, since the blowout began.

On June 10 the official government-accepted rate of leakage gets doubled. The U.S. government's flow rate assessment team announces, "The lowest estimate that we're seeing that the scientists think is credible is probably about 20,000 barrels, and the highest is probably a little over 40,000." Twenty-five thousand, near the low-end estimate, is over 1 million gallons a day. The *Exxon Valdez* tanker leaked an estimated 11 million gallons.

"I think we're still dealing with the flow estimate. We're still trying to refine those numbers," says—guess who—Coast Guard admiral Thad Allen. Almost certainly the rate is changing; it may be increasing, because once oil starts flowing out of a geologic formation, the rock erodes with the flow and the channels enlarge.

The Gulf isn't the only thing hemorrhaging; BP stock closes down 6.7 percent, hitting its lowest level since 1997. The company's market value has spilled billions. Its share price has collapsed more than 40 percent since the blowout began, leading some to raise the possibility of bankruptcy.

It occurs to me that this would be a time to buy, if I hadn't sworn off fossil fuel stocks. Bankruptcy is just wishful thinking. BP is the third-largest oil company in the world, after ExxonMobil and Royal Dutch Shell, with 80,000 employees, sales of $239 billion in 2009, and a market value—even after the recent losses—of more than $100 billion. BP is multinational, traded on both the New York and the London stock exchanges, with Brits and Americans on its board of directors, and extensive U.S. holdings. In 1998 it merged with the American oil company Amoco. About 40 percent of its shares are held by American investors. Its Texas City refinery is one of the world's largest, and BP owns 50 percent of the Trans-Alaska Pipeline.

Sarah Palin calls BP a "foreign company" because, well, she's a little behind in her current events. The White House knows better, but it, too, is whipping up anti-foreign sentiment by consistently calling BP by its former name, British Petroleum. And so in Britain—where BP is, in fact, an evocation of the glory of the empire, a huge tax contributor, and thus beloved—Conservative peer Lord Tebbit calls the American response "a crude, bigoted, xenophobic display of partisan, political, presidential petulance." He may think America's response is much too crude, but from our vantage, BP has provided America with too much crude.

The fisheries closures continue expanding. Now totaling 88,522 square miles. About 37 percent of the Gulf's federal waters. Federal waters begin three miles from shore, but most state waters are also closed. More than half the Gulf remains open to fishing, but buyers are canceling orders. "I've had guys saying, 'If it's from the Gulf, we don't want it,'" says a New York City seafood distributor. The celebrity chef says, "People are really wondering if we're getting safe fish." In Chicago and elsewhere, restaurants display signs declaring, "Our Seafood Is Not from the Gulf of Mexico." "They believe it's toxic," a New York chef says. "So let me be clear," says the president of the United States. "Seafood from the Gulf today is safe to eat." The New Orleans sales rep who ships fish nationwide says, "They're not ordering anything. Not a one. They know we're not selling tainted fish. But their customers? No way. They don't want seafood from Louisiana at all."

"Everybody is so stressed here. We're just sitting here waiting and they're not telling us anything because they don't know," says a Grand Isle restaurant owner who may soon be out of business. "I had four people who came yesterday crying." A fisherman says, "My wife cried and cried over this. Just the other night she told me, 'Thank God there isn't a loaded gun in this house.'"

In Gulf Shores, Alabama, thick oil washes up at a state park, coating the white sand with a thick, red stew. "This makes me sick," says one resident, her legs and feet streaked with crude. "I've gone from owning a piece of paradise to owning a toxic waste dump." Says a fishing guide, "I don't want to say heartbreaking, because that's been said. It's a nightmare. It looks like it's going to be wave after wave of it and nobody can stop it."

Meanwhile, dozens of oil-drenched pelicans float around Louisiana's Grand Terre Island. People have found more than 500 tarred-and-feathered birds dead, and have rescued about 80 oiled birds and nearly 30 mammals, including dolphins. Most showed no obvious oil; maybe something else killed them. But oil ingestion and fumes could have caused this.

The sight of animals struggling in oil moves me to tears more than once. But the numbers here are small compared with the avian toll of the *Exxon Valdez.* After that spill, workers immediately found more than 35,000 birds; by reasonable estimates, approximately 250,000 died. That was because of the density of the oil, the temperature of the water, and the fact that coastal Alaska is home to enormous numbers of aquatic birds of whole family types that (like sea lions) don't live in the Gulf.

That doesn't mean the rescue efforts are going well.

"This is the worst screwed-up response I've ever been on," says Rebecca Dmytryk, who has worked with oiled birds in Louisiana, California, and Ecuador, and founded a group called WildRescue. Experienced wildlife rescuers have complained that they've been prevented from going out to look for live oiled birds in the most likely places, sidelined, or never called in at all. A guy with the U.S. Fish and Wildlife Service says that rescuing oiled birds is a task for "our trained biologists." But experienced rescuers complain that the job of rescue went to inexperienced government employees—fisheries biologists, firefighters—who had never touched a bird before. "I'm just at a loss for why this was allowed," says Lee Fox of Save Our Seabirds, who has written a manual on handling oiled birds.

Jay Holcomb of the International Bird Rescue Research Center has been saving birds from oil spills for thirty years, on three continents. During the *Exxon Valdez* event, he oversaw the entire bird search and rescue program in Prince William Sound, the largest ever attempted, involving dozens of boats and thousands of birds. But here, that's not good enough for the officious officials. "We've been assigned to take hotline calls," he complains, "completely kept out of it."

BP hired a four-year-old Texas company called Wildlife Response Services to oversee the rescue and rehabilitation of birds, turtles, and any other animals hurt by the spill. Its owner says Holcomb and the other wildlife rehabilitation experts "didn't have the personnel to go out and rescue all the birds." She says the system she set up has worked well, adding, "I don't know why anyone would question that."

Four months ago, she did call Lee Fox and tell her to get ready. Fox says, "I've never heard another word. I'm up to my nostrils drowning in frustration." Dmytryk says she and her coworkers begged for permission to go out into the Gulf to look for sick and injured birds that were too weak to make it to shore. But they were turned down. Sharon Schmalz of Wildlife Rehab and Education in Texas, who has over twenty-five years' experience working spills Gulf-wide, says, "We were told to stay put." "They said for safety reasons we couldn't do it," Holcomb says. "There was not a lot of interest in using our expertise."

Meanwhile, a blowout in Pennsylvania: a well blows natural gas and drilling fluid seventy-five feet into the air. It does not ignite and no one is hurt but it takes sixteen hours to control.

The government now estimates that 500,000 to 1 million gallons of crude—12,000 to 24,000 barrels—are leaking daily.

Our Thadmiral tells us, "This spill is just aggregated over a 200-mile radius around the wellbore, where it's leaking right now, and it's not a monolithic spill. It's an insidious war, because it's attacking, you know, four states one at a time, and it comes from different directions depending on the weather." He adds with a dash of frustration, "This spill is keeping everybody hostage."

Hostages: A BP rep tells residents gathered at a church, "We are all angry and frustrated. Feel free tonight to let me see that anger."

Residents aren't buying it. "'Sorry' doesn't pay the bills," says one. "We're sick and tired of being sick and tired."

Sick. And tired. In Louisiana, seventy-one people suffer throat irritation, cough, shortness of breath, eye irritation, nausea, chest pain, and headaches following exposure to emulsified oil and dispersant. Most are briefly hospitalized.

The stress of anger is giving way to the hopelessness of depression. Without fishing, what's lost is not just vocation, but also what life means, what life is, and people's understanding of who they are. What's lost is pride. What's gained is fear of losing everything. What's creeping in around the edges: The search for answers at the bottom of a bottle. The thought that suicide may end the pain.

In two weeks spanning the last week of May and the first week of June the Louisiana Department of Health and Hospitals counseled 749 people having symptoms that could lead to destructive behavior. Experts say the region should brace for long-term psychological strain. Are they making matters worse by announcing that?

One fisherman, stricken by the sight of fish floating dead, frets over whether he will be able to pass on his trade to his children, a thirteen-month-old son and ten-year-old daughter. His wife, who has sought counseling, says, "My husband went from a happy guy to a zombie consumed by the oil spill." He replies, "If you're not out there in it, you can't comprehend what this is about. We're going to be surrounded by it." Says one town council member, her voice trembling, "We're not going to be okay for a long, long time."

HIGH JUNE

The flow BP is getting good at stopping is the flow of news. When folks at Southern Seaplane, in Belle Chasse, Louisiana, call the local Coast Guard–Federal Aviation Administration command center for routine permission to fly a photographer from the *Times-Picayune* over part of the oily Gulf, a BP contractor answers the phone. His swift and absolute response: Permission denied. "We were questioned extensively. Who was on the aircraft? Who did they work for?" recalls Rhonda Panepinto, who co-owns Southern Seaplane with her husband, Lyle. "The minute we mentioned media, the answer was: 'Not allowed.'"

A spokeswoman for the Federal Aviation Administration says the BP contractor who answered the phone was there because the FAA operations center is in one of BP's buildings. "That person was not making decisions about whether aircraft are allowed to enter the airspace," the spokeswoman spoke.

Why is the FAA in a BP building when BP is the cause, and is under criminal investigation? No other office rental spaces in the four-state region? And they're sharing phone lines?

Across the Gulf, I as well as various photographers, journalists, filmmakers, and environmentalists trying to understand and document the spreading oil are now having real problems. We're getting turned away from public areas affected by the oil—and being threatened with arrest—by private guards, sheriffs, cops, and the Coast Guard.

Senator Bill Nelson, Democrat of Florida, planned to bring a small group of journalists with him on a trip he was taking through the Gulf on a Coast Guard vessel. The Coast Guard agreed to accommodate the reporters and photographers. At about 10:00 P.M. on the night before the trip, someone from the *Department of Homeland Security* called the senator's office to say that no journalists would be allowed.

What lame excuse did they have? "They said it was the Department of Homeland Security's response-wide policy not to allow elected officials and media on the same 'federal asset,'" said a spokesman for the senator. "No further elaboration."

A reporter and photographer from New York's *Daily News* were told by a BP contractor that they could not access a public beach on Grand Isle, Louisiana, one of the areas most heavily affected by the oil spill. The contractor summoned a local sheriff, who then told the reporter, Matthew Lysiak, that news media persons had to fill out paperwork and then be escorted by a BP official to get access to the public beach. "For the police to tell me I needed to sign paperwork with BP to go to a public beach?" Lysiak said. "It's just irrational."

BP is obviously a company with a lot to hide. But how it's staged a coup of the Gulf and gained control of government—*that,* I don't get.

"Our general approach throughout this response," says yet another of the seemingly dozens of faceless BP spokesmen, "has been to allow as much access as possible to media and other parties without compromising the work we are engaged on or the safety of those to whom we give access."

They *allow?* To whom *they* give access? How did a corporation succeed in suppressing U.S. citizens trying to see and talk about what's going on, and *why* are any of our law enforcers, who should be guarding the coast against BP, so thoroughly and sickeningly capitulating, deferring, and letting themselves Be Played? Obama wanted to know "whose ass to kick"? The answer's so blindingly obvious. How is that even a question?

When CBS News reports that one of its news crews was threatened with arrest for trying to film a public beach where oil had washed ashore, the Coast Guard says it is disappointed to learn of the incident.

Signs announce imaginary lines, but the real landscape changes slowly from Louisiana to Mississippi to Alabama to Florida. And so does the light. Green fields. Blue skies. Black cows. Red barns. Sentinal mockingbirds. Amber waves of wheat. Hay for sale while the sun shines. Modest houses with wide lawns and the shade of big trees. Chairs on porches. Mimosas in bloom. Spreading palms, tall pines. The proudness of corn. A few unlucky armadillos. The Roadkill Cafe, the Elberta Social Club. Antiques and collectibles. Live bait and crawfish at the hardware store and at the grocery store. At the garden store: "Ten Percent Off All Firearms." The tank guarding the veterans memorial. Signs directing our attention to pizza, the control of pests, and eternal salvation. A Baptist church advises, "Do Your Work Today As If There Is No Tomorrow." Dopey advice; all work is about tomorrow. Tomorrow, opposable thumbs, and the ability to ignite oil are what make us human.

But for fishermen and anyone dependent on tourist dollars, perhaps yes, now's not a time to wreck your head over tomorrow; too many unknowns. Take it one day at a time. On the other hand, to get you through this, the most important thing is to keep imagining a better tomorrow. Another church and a sign a bit more to the point: "Forgive Us, Lord." That covers the multitude of sins, serves up the Big Prayer.

Nighthawks and chimney swifts. Small bridges. Their rivers and bays lined with emerald summer. Troops of pelicans flying west. I've come from the west. West is where the Oil is coming from. It dominates even my sense of direction. Coats my mental compass.

The scenery, changing slowly. Tractors for sale. Tractors at work. Signs advertise the services of those who weld, those who build docks. "Deep Sea Fishing: 4.4 miles." Close by but already in the past. A man named Twinkle is running for office.

One of life's simple pleasures: driving with the radio on. I hear that the wife of one of the eleven killed says BP will never feel the pain the survivors feel. But how could it? It is not a person. Where a heart would be, it has only money.

The Supreme Court disagrees with me; five of its justices say a corporation *is* a person. Does a corporation have a belly button? That's one's passport down through ages, a living link in the one eternal chain of being, life to life. BP is no person. Person: a two-legged primate with thoughts, feelings, blood ties, dreams. Not something for courts to trifle with. Not, in truth, subject to their opinion. A sacred matter, of the greatest of mysteries.

First glimpse of Mobile, Alabama, includes too many oil rigs to count easily. One reason I've so seldom returned to the Gulf is the rigs' visual blight. They're the wide horizon's piercing reminder of oil's stranglehold on us, our cheap-energy addiction. The rigs flare their gas from their long arms like Lady Liberty's torches. We are all victims, all perpetrators. What goes around comes apart.

The shorelines of Mobile Bay are confettied with orange boom. It's hot. The sun raises an ocean haze. Right now here, the Oil is mostly an idea, the coming thing. The booms await it. The rigs foreshadow it.

On the corner, the BP station. I still have half a tank.

Scientists continue to plague us with pesky reports of massive plumes of undersea oil. Woods Hole Oceanographic Institution researchers working aboard the National Science Foundation's ship *Endeavor* beginning in mid-June confirm the presence of drifting oil deep below the surface. Using 5,800 separate measurements, they find a hydrocarbon plume over a mile wide, over twenty miles long, hundreds of feet thick, detectable down to 3,300 feet, and extending itself at a little over four miles a day.

People envision a river of oil. But Woods Hole chemist Chris Reddy will explain, "The plume was not a river of Hershey's syrup." The water samples collected at these depths had no odor of oil and were clear. The dilute oil was detected by instruments. "But that's not to say it isn't harmful to the environment," he adds.

And in perhaps the world's first case of plume envy—or whatever you'd call it—the University of Georgia's Samantha Joye says her plume is bigger than theirs; she says Woods Hole's plume "doesn't hold a candle to the plume we saw."

Whether like syrup or seltzer, the thought of massive plumes of undersea oil is disturbing. So from what corner shall come the next volley of fresh reassurance? National Oceanic and Atmospheric Administration (NOAA) officials call the academic scientists' announcements of their discovery of plumes "misleading, premature and, in some cases, inaccurate."

Note: "in some cases." Interesting phrase. An exception big enough for a wiggly pig. Means—it would seem—that "in some *other* cases" it *is* accurate to say there are giant undersea plumes of hydrocarbons now drifting in the Gulf.

NOAA says oxygen depletion in the waters surrounding plumes is not "a source of concern at this time." NOAA says critics blaming dispersants for the plumes have "no information" to back their claims up. NOAA administrator Dr. Jane Lubchenco, whom *Newsweek*—in a carefully crafted phrase—calls "a respected oceanographer when President Obama tapped her to lead the agency," says there are no "plumes," only "anomalies."

Now, wait a minute. Don't tell me "no information." Obviously, the oil is coming out of the seafloor. The dispersants are designed to keep it underwater. It's *coming* from beneath the surface. It's *being dispersed* beneath the surface. And when it surfaces, it's being hit with dispersants that dissolve it to make it *sink.* Beneath the surface is where a lot of it is—obviously.

I've known Jane Lubchenco for close to twenty years, and in the late 1990s I spent a lot of time with her when we were part of a team traveling to various cities to tell people about how the ocean is changing. I respect her very much. But there comes a time when political pressures cause a person to try to distinguish between "plumes" and "anomalies," though those are two words for the same thing. One can parse language into droplets so small that the facts dissolve to difficult-to-detect dilutions; one can disperse truth itself. People get confused, get the wrong impression. Sometimes that's the goal. So it's better to stand with the truth. Political situations come and go. What begins and what ends, and what follows a person forever after, is the truth. Obviously the oil is in the water. It comes from somewhere, it has to go somewhere, and the water is where.

A scientist at Columbia University's Lamont-Doherty Earth Observatory says that even without dispersants, oil originating with such force at this depth and pressure breaks into zillions of droplets that stay suspended in the water.

Not as visually shocking as dead pelicans, but much more basic, is the plight of the minute clouds of life at the base of the food chain. The things that grow the fish that magically become leaping dolphins and plunging seabirds—tiny things living deep, almost beyond the reach of human acknowledgment—are probably having a pretty hard time right now across large swaths of the Gulf.

The deep sea contains a galaxy of little animals—everything from planktonic beasts to jellies to billions of small fishes called lanternfishes (myctophids). They all live in a layer of life that twice a day performs the greatest animal migration on Earth, from the darkness of deep daytime waters to the darkness of shallow nighttime waters, and back again. This zone of life is like a flying carpet throughout an astonishingly large portion of the world ocean. On a moving ship, it is extraordinary to watch it on the sonar as it slowly rises and dives over the course of the day, even while your ship is covering hundreds of miles. Proponents of dispersants contend that by dissolving the oil, microbes can more easily feed on it. But before that happens, these billowing toxic clouds will roll through this zone of life in the Gulf, where the damage will likely never really be assessed.

I'm not saying it's utter catastrophe for all of them. I'm not saying they won't bounce back. I'm saying, Let's not fool each other. Let's attend to the matters to which the researchers are calling our attention. Let's not be in denial of science, logic, and sense. That's indecent.

BP's CEO, Tony Hayward, insists there's "no evidence" of hydrocarbon plumes in the Gulf. There are wiggly pigs, and then there are liars.

For an alternate take, we turn to Florida State University oceanographer Ian MacDonald: "These are huge volumes of oil, in many cubic kilometers of water."

The National Oceanographic and Atmospheric Administration acknowledges that it has "confirmed the presence of very low concentrations of sub-surface oil at depths from 50 meters to 1,400 meters." In other words, from shallow waters down to around 4,000 feet. But they say it's in "very low concentrations," less than 0.5 parts per million. The thing is, a concentration of hydrocarbons in the range of less than 0.5 parts per million—say 0.4—is in the range of a notably polluted place like Boston Harbor. So one person could say, "It's not going to kill everything," and another could say, "The oil is polluting an enormous volume of water in the Gulf of Mexico." They'd both be right.

From the federal government, we get both a little more minimizing—"We have always known there is oil under the surface"—and this honest admission of ignorance: "The questions we are exploring are where is it, in what concentrations, where is it going, and what are the consequences for the health of the marine environment?" So asks NOAA administrator Dr. Jane Lubchenco.

Those are the questions. If the oil stays suspended it will eventually dissipate throughout the world ocean in rather harmless concentrations—the dose makes the poison—or be dismembered by microbes. Another possibility is that the dispersed oil might sink to the seafloor. A fisherman I know on the West Coast e-mails me: "During the 1969 Santa Barbara blowout dispersants were used extensively. Several years later, some of the oil that had settled to the bottom eventually formed large clumps. In the early/mid 1970s I was working on drag boats in the SB Channel and we would periodically hit one of the clumps, which would render the net completely worthless. No way to clean that shit off the net. Union Oil wound up paying for a lot of nets, although not willingly. No one knows the effects on bottom life, although the English sole disappeared for years."

At a press conference, Lubchenco, sounding like a bomb-sniffing canine working in a political minefield, says, "The bottom line is that yes, there is oil in the water column. It's at very low concentrations." That sounds like she's minimizing it. But in her next breath she adds, "That doesn't mean that it does not have significant impact."

Taken together, a true enough picture. Depends on your definition of "significant," but I'm satisfied. Even if she didn't hit a home run, she touched the three bases of truth in this mess: it's there; it's at low concentrations except near the source; and it could still be a problem. I think that's really as much as anyone knows. This isn't a home-run situation.

Dauphin Island, Alabama. A pretty place. A mixture of fishing village and resort. A beach place with pines tall enough to cast shade across the road, a place that welcomes you with a sign proclaiming, "Fishing, Beaches, Bird Sanctuary." The Oil welcomes you with a sign saying, "BP Claims Center."

It's June 8; today is World Oceans Day. I'm having a bit of a hard time, emotionally speaking, with that.

Away from the pines and alongside the sea oats, by 10:00 A.M. the sun is uncomfortably hot, its glare unrelenting, resolute.

I'm with filmmaker Bill Mills, a real pro who's done a lot of work with *National Geographic* films and many others. He wants to interview a few people, and he's toting his movie camera as we walk from the parking lot toward the water. Someone in the shade of one of BP's little sun awnings starts yelling at us. Bill doesn't even turn around. I glance over my shoulder. An arm-waving man, too lazy to come our way, demanding we come his way. Public beach, so screw him.

A light, ribbony slick gift-wraps the shoreline like a Big Present. A few hundred people are still putting a brave face on beachgoing. Determined to have their beach day. In their swimsuits. On their blankets. The sun still works for them. Admirable fortitude.

Not surprised to see oil globs splattering the beach. Surprised to see some kids in the water.

Didn't see the sign that said no swimming? It's easy to miss: tiny, smaller than a stop sign. "The Public Is Advised Not to Swim in These Waters Due to the Presence of Oil-Related Chemicals."

Just a small cover-your-ass sign in a community reluctant to admit that it's ruined. Ruined at least for now. It could be much worse. Just go to Louisiana's beaches. But what's arrived here now is quite enough to spoil things.

A mom: "We come here a *lot.* It's *killin'* me to see this."

BP workers in white suits rummage among the beachgoers like foraging chickens pecking at oil. Near the tide line they shovel oil blobs and oil-stained sand into plastic bags.

An old man says he was a newscaster for forty years, that's enough of being on camera, so no photos, please. Now he casts a fishing line. "Other day I went fishing over there, where I've fished for thirty-five years. The place was full of tar balls and the fish I caught smelled like gasoline. I *just had to leave!*"

Tar balls I believe; there's plenty. A fish that smells like gasoline? I'd have to smell that myself. Healthy skepticism is how I like to think of it. It means that, at least sometimes, I'm seeing through my anger.

Ashore, a wet boy, nineish, looks bewildered. Won't go back in. "You feel it all over your skin," he says. Another family's kids are body-boarding, looking perfectly content.

Other creatures continue to work their world, their side of things. Least terns diving along the shore. The small fish. A school of mullet nose the surface. A small ray swims past: another hazard to swimmers. It's a very alive place. That's the point. The people and the others that live here. Everything that lives here.

A man emerges. I ask him if he can feel and taste oil. He says, "Oh, I can feel it on me. But as far as tastin' it or anything, no, nothin' like that."

Mike, thirty-nine, wears American flag swim trunks. He stands with his well-tattooed arms folded. Feet oil-stained. "We come with the family on weekends. I'm worried this is the last time. Forget this oil; go solar power."

You can clean a beach, but every wave brings a little more. The oil is still gushing, and there's plenty more to come. Every day, oily sand goes out in bags. Bags and bags. Bag People.

The BP workers wander near. I ask what kind of work they would be doing if not for the Oil. "I have no comment," they regurgitate.

"They tell you not to talk to anyone?"

"Yes, that's right."

The Oil is a big secret. Okay, I wouldn't want my employees talking to anyone while on the job, either. But there's a motive here: minimize the scope of information and opinion. Aren't we really quite past the point where that can help even BP? What might these people say that we don't already know? That the genie's blown his cork, Pandora's opened the coop, and the cats are out of their bags all over the region? Basic Privacy is no longer this company's prerogative, and the pettiness does less to hide the facts than it does to expose the corporation's desire to hide facts.

In a cloudless sky, a Coast Guard helicopter passes.

Outside one of the local churches, I meet Reverend Chris Schansberg. He's forty-three, soft-spoken. "The first Sunday after the explosion," he says, "I told our people, 'Let's get together and pray that the oil won't hit the island.' I don't know why we focused on our island, but we did. And I was talking to the mayor later, and he didn't know we'd prayed, and he said, 'Y'know, it's really remarkable that oil hasn't hit the island.' I know it hit the far west end. But on the oil-flow charts on the evening news, people were saying, 'Now, look at Dauphin Island; there's a seven-mile buffer between it and the oil.' And apparently—to the glory of God—oil's been flowing around us. Now, that's a great victory. But of course, you have to answer, Well, why is it on the shores from Louisiana to Pensacola? That's a deeper question." Brief pause. "The oil cleanup workers come from everywhere, from Washington to New York State, Puerto Rico. I've seen loneliness, isolation, boredom—. Many of them have never seen heat like this. We have seen this as an opportunity to reach out toward them, sharing God's love, giving out frozen Popsicles—.

"Someday, if not now, the people who've done this will be judged. I'm not saying it's unforgivable. But they *are* accountable to God. If they tried to cut corners, if they failed in their stewardship, they will be accountable on the Day of Judgment. The Bible promises that when Jesus returns he will make all things right. And it really means everything. So if we still have oil in the Gulf when he returns—that's it."

Fifty House Democrats begin calling for BP to suspend its planned dividend payout, stop its advertising campaign, and instead spend the money on cleaning up the ongoing Gulf of Mexico oil spill. Congressional Representative Lois Capps says, "Not a single cent" should be spent on television ads. If BP is so concerned about its public image, "it should plug the hole." Yes, but—. Suspending a dividend payment would save BP something like $10 billion. The loss of prestige and stock value is a strong deterrent for not paying a dividend. But for BP, the silver lining would be pretty convenient cost savings. And it could simultaneously deflect its shareholders' anger over plummeting stock value by blaming the mean old U.S. congressional Democrats.

The depth of 2009's global recession sent global energy consumption down for the first time in twenty-seven years, but only by 1 percent. In the same year, China's carbon dioxide emissions from fossil fuels rose by 9 percent. Global wind and solar generation rose by about 30 percent and 45 percent, respectively, enormous gains. Who says? A BP energy report says. The world's oil reserves (including yet-to-be-developed Canadian "tar sands") appear sufficient to last—at 2009 production rates—for forty-five years; gas will last about sixty-five years, and coal will last 120 years. After that, what? Do you or does someone you love plan to be alive forty-five years from now? This is our wake-up call. Don't touch that snooze button.

By mid-June the cap is ferrying about 10,000 barrels of oil a day up to a tanker on the surface. The well is leaking much more than that.

Revisiting the basic question: How much oil? A team of researchers and government officials is finally studying the flow rate. Team member and Purdue University professor Steve Wereley says it's probably somewhere between 798,000 and 1.8 million gallons—roughly 20,000 to 40,000 barrels—daily. He says, "BP is claiming they're capturing the majority of the flow, which I think is going to be proven wrong in short order."

Meanwhile, at the start of the second week of June, the tide begins leaving a splattering of oil blobs on Florida Panhandle beaches, from Perdido to Pensacola. Signs warn: Don't swim. Don't wade. Avoid skin contact with oily water. Avoid dead sea animals. Notes one: "Young children, pregnant women, people with compromised immune systems and individuals with underlying respiratory conditions should avoid the area."

Some people are in the water anyway. I wouldn't want to get in where I can see a sheen, but it's probably not very toxic in small amounts for short intervals.

Tourism, though, has been completely poisoned. In Mississippi, Governor Haley Barbour angrily blames the news media for scaring away tourists and making it seem as if "the whole coast from Florida to Texas is ankle-deep in oil."

The fact is, he's right; it isn't the same everywhere. As Thad Allen intones, "We're dealing with an aggregation of hundreds of thousands of patches of oil that are going a lot of different directions. It's the breadth and complexity of the disaggregation of the oil that is now posing the greatest clean-up challenge." And so, some places are ankle-deep and some, barely freckled.

Pensacola Beach, Santa Rosa Island, Florida. The sand here makes the "white" sand back home on Long Island seem battleship gray. In bright sun it can be blinding. Workers have picked up the day's dark tar balls. Mostly.

High-rise hotels and condos loom over the low-slung beach. They shimmer in the sunset. A huge restaurant with a huge fake crab is overbuilt and tacky. The whole place is not my cup of tea. But it's not *my* home.

Listen to people whose home it is: "So, this is Santa Rosa Sound," she says, as if presenting it as a gift. "Really precious." Then, almost under her breath: "I don't know, it may be twenty years before it ever looks like this again. The oil's just a few miles offshore. Oh look—a dolphin." A moment's silence, then: "There's no plan to help the dolphins."

The evening breeze is still warm enough to raise a sweat. The clouds paint with pastels. A guy with a floppy hat, a waddling walk, arrives, sighs, "I'd like to take one last look at my heaven."

BP's full-page ad promises to invalidate their fears. "We Will Make This Right," it announces, projecting confidence. But just below the horizon, flying under the radar, on everyone's minds, on their nerves: oil blobs stalk.

At public faucets, moms rinse sand from kids' toes. News vans, their satellite dishes turned to heaven as if in prayer, stand like military hardware on the eve of invasion. Ready to report attack. We breathe the parked vans' continuous exhaust, their engines running their air conditioners and electronics.

A Big Protest is scheduled for here, now, this moment. But so sparse are the protesters—fewer than twenty, among several hundred beachgoers—that even the organizers cannot locate them.

Finally, the protesters. There they are. Here they come. They have signs—"Save the Dolphins"—and a little theater; a woman wears a black veil of mourning. They group up and knock on a media van. There're enough of them to fill a frame. Passable for local news. Dutifully reported. Story filed: "Local People Do Not Like Oil on Beaches." Basically Pointless.

One police car stands idling, its occupant utterly relaxed, a bit bored. The woman beneath the veil says, "We already got the *police* here—and I ain't even showed my ass yet."

Oh boy. Time to go.

Restaurant. "The oysters are from—?"

"Texas," says the waiter. "We might have a month left there."

Realtor, late forties, energetic blonde, clear blue eyes, too much Sunshine State sun, a tough cookie: "People come for the beaches. If they can't put their feet in sand, they won't come." She's already refunded 75 percent of this season's rental down payments. Her home sales are in cardiac arrest.

Activist folksinger to me: "Do you think we can save the dolphins?"

"If it gets down to them needing to be saved? No."

"There should have been a plan." She wipes her face, gazes out toward the dock. "It's weird to fight as you're grieving the loss of your own place. With global warming, you feel some distance. But this. When the rig sank on Earth Day, I cried myself to sleep. We take turns picking each other up."

Others chime in. "If the wind tonight was off the Gulf like last night, we couldn't sit here. The smell was thick. We've all been exposed."

"People are sick. More people are gonna get sick. I took two reporters out and they both got sick. One was in the hospital. Karen right there's got chemical burns from picking up four dead turtles."

"This is war. This is all-out war. This is a story that has to be told. And a powerful, very *powerful* group does not want this story told. I'm just me. I've been helping where I can. But my phone's having problems. My sister's been helping, too, and her e-mails are getting kicked back to her. I'm not paranoid, but what if somebody's interfering with our communications? Do you know how I could find out? Since the day the president came through, my phone really hasn't worked very well. I don't know. But that's what's happened."

"Every time the oil moves, they change the no-fly zone. They don't want us to see."

"How many notebooks do you fill up before you have a book?"

"Karen's trying to get her kids out of town. There's nothing for them to do. Everything we do is sailin', crabbin', boatin', shrimpin', surfin'."

"I'm really worried about Lori; these dolphins are like her children."

"If a kid says the water tasted soapy, is that dispersant?"

This is the conversation now.

But while we talk, the water still looks fine. The pelicans unruffled. Gulls galore. Dolphins roll within a hundred yards of the restaurant deck. All looks safe. Nothing feels safe.

The real estate agent says, "If this keeps up much longer, we're dead. I have people who, if they can't rent their houses, they'll get foreclosed. They don't have the money to refund deposits. And they can't sell; who's gonna buy anything now? Everybody's hit. The gas stations. The laundry services. Even the hairdressers depend on tourists. Gonna be a ghost town."

Another volley of comments from around the table. "I never thought this could happen."

"I'm stayin' to the end. Till they make me leave. This is my home."

"This was my *dream.* I'm *really* mad. I'm upset enough to have dessert."

On the futility and utility of helping wildilfe, a debate: More than a thousand birds in the Gulf region have been collected alive with visible oil. Serious question: Should they be cleaned or killed? Two years after a 1990 spill in southern California, fewer than 10 percent of oiled brown pelicans that had been cleaned remained alive, and they showed no signs of breeding. "If nothing else, we're morally obligated to save birds that seem to be savable," says one bird worker. Several hundred Gulf sea turtles are also getting aid. But a University of California professor who worked to save animals after the *Exxon Valdez* spill in 1989 says cleaning wildlife gives a false impression that something can be done. To which the director of the International Bird Rescue Research Center retorts, "What do you want us to do? Let them die?"

In mid-June, another vocabulary word hits Alabama: "mousse." Think of it as crude oil the consistency of chocolate pudding. But remember, it's not pudding. If you step in it, you can't wash it off. You need some kind of solvent. If it's just a little, you can maybe rub until you've gotten most of it. If it's on clothes, throw them away. If your kids or dog track this stuff into the house, you need a new rug or maybe a new couch.

On June 14, 300 birds get coated with oil—in Utah.

Car radio: ". . . We've just been hearing from Riki Ott, author of *Not One Drop: Betrayal and Courage in the Wake of the* Exxon Valdez *Oil Spill*. She was heavily involved in the *Exxon Valdez* disaster. If you have any questions for Riki Ott about what's going on in the Gulf, give us a call. . . ."

I once met Riki Ott; I've been in her home, actually. She lives in Cordova, Alaska. Former salmon fisherman with a PhD. Unusual woman.

"Riki, you've been saying—"

"In this country, we've made the decision to depend on the oil industry for cleanup. We put the spiller in charge. In Norway, the government nationalizes the spill; they have the equipment, they do the cleanup. In this country, we put the irresponsible party in charge."

"So we say, 'You're irresponsible, now you're in charge—' "

"Right. We've been trying to get the workers respirators. Something as simple as respirators. We need federal air-quality monitoring. Because if we rely on BP's monitoring, it will end up where Exxon's air-monitoring data ended up: disappeared—sealed into court records until 2023."

"You were saying earlier that there's some trouble with just getting to the sites."

"The oil companies learned a heck of a lot more than the citizens in the wake of the *Exxon Valdez*. And what the oil companies learned is this: control the images. No cameras. No evidence. No problem—right?"

A royal tern and a brown pelican watch Captain Cody McCurdy pull the *Gray Ghost* from its pilings at Orange Beach, Alabama. Tommy Gillespie, mate, wears old tattoos on weather-beaten arms, a Confederate bandanna, two packs of cigarettes in his T-shirt. Fishermen. Until last week or so.

"I don't see a way out," Cody says, but he means it literally.

The marine police are enclosing the inlet with boom. Various boats stand by in case they're needed to deploy more boom. There's not enough for all to do. The Oil inflicts a certain aimlessness.

All boats, many of them recreational boats, are now "workboats." BP's money flows like oil. BP's managers can't turn *it* off, either, or they'll trigger another eruption: people's anger. They know that to keep people Busy and Paid—no matter how useless the errand—serves their bigger purpose: to quell, to calm, to keep the masses anything but idle. The people are Being Pacified. This busywork is theater. And to that extent, BP is now the Gulf's biggest patron of the arts. The obvious calculation: paying for theater to suppress rage is well worth it.

The folks on the boats don't see themselves as being manipulated. They are desperate for the income, and it's enough that they're getting what they've Been Promised. They can't afford to look behind that curtain. Many feel—deeply—that they are defending their home waters and wetlands, defending the most important and meaningful thing in their lives.

While we wait to see if we can get out of here, I notice that everyone aboard all the other boats wears a silly little orange life jacket, the uniform of Being Paid, even in water calm enough to reflect one's tightening anger. Most professional fishermen have probably never in their lives worn a life jacket on a boat. But of course BP wants to ensure *safety on the job.* Unlike while drilling in mile-deep water amid risks galore.

Even ashore, I see fishermen walking the docks—their *own* docks—in their little orange life jackets. Part of how Being Paid wipes away their autonomy, their adulthood, their discretion, their individuality; infantilizes them.

Fishermen are famously talkative, but now the vest wearers "have no comment." I pity their sorry slavery. But I can afford my anger; I can go home.

We are the *only* boat on the water not drawing a BP check. We wear no orange vests; we have signed no waivers. I can abstain indefinitely.

My companions, their options narrowed to the Breaking Point, can't. They're going to lose their bet that the gusher will end soon enough for them to salvage their sense of self. They just got their hazmat training and say they're hoping to start working the oil cleanup response soon.

Can't blame 'em; everyone's under Big Pressure to make a day's pay.

On the tide in the pass: lumps of oil.

Basic Problem: The marine police, blue lights flashing, say that if we go out, we won't be allowed back in.

Cody says his boss knows the cops. His boss makes some calls and rings back, saying, "They aren't letting anyone back in here. They don't know why."

A well-planned, well-coordinated, well-executed plan is all that's lacking. Speculation remains open. Reason is closed. The officious are expert at Being Petty.

Cody says screw it, we'll go. "Open up."

We's takin' our chances, betting on the unreliability of the marine police's advice, on inconsistency, on lack of coordination. Those guys may be off their shift by the time we get back. The next guys may say something else. And maybe it's easier to say no to a boat leaving than to one coming back? We'll see. They all seem to be making it up as they go.

One thing's certain: opening a line of boom to let a boat back in makes zero difference as far as the Oil is concerned. This boom is useless against it. You might as well stretch dental floss across your bathtub to hold soapy water to one side.

They lower the boom on our wake. So much effort, applied so diligently, so earnestly—and so unequal to the task at hand.

At 9:00 A.M. the sun squints my eyes. How do the locals stand such heat?

Our mission: find the oil before it finds us. Get a sense of where. How far. How heavy. Paul Revere with outboard power.

Pelicans and helicopters. Ospreys and helicopters. Royal terns and Coast Guard helicopters.

Looking back from three miles off. High-rise condos line the shore. Ugly as hell. No boats out here, just helicopters. All fishing, even catch-and-release, is closed. A fishing boat may not possess fishing gear. That's to *protect* us.

Safe from fishermen, the fish must fend for themselves in the dispersant-and-oil seawater. In the low-oxygen end of the pool. For them it's always either frying pan or fire.

We head to a great fishing spot fifteen miles from shore. Rumors are: someone saw dead fish at the surface there. We are fishing for rumors, fishing for dead fish.

There's enough wind for a light chop, but the water is calm. Too calm. A bit slick. An ocean normally shows many natural slicks, caused by the bodily oils from schools of fishes, by water of different densities—nothing to do with petroleum. I ask Cody if he thinks these slicks are natural. He says, "What slicks?" I point.

"Naw, this is just—" He waves it away.

From the back of the boat Tommy yells, "A lot of oil here."

We're suddenly sliding past coin-sized oil blobs. By the thousands.

Then, none.

Schools of tuna-cousins called little tunny shred the surface chasing small-fish prey. A couple of porpoises, far off. Blue sky. Blue-green sea. A light whiff of oil. Not as bad as I've heard tell.

When I ask Cody where he fishes—fished—for tuna and marlin, he points left and says seventy to ninety miles thataway. The big fishes' address is the names of three oil rigs way out there. The rigs float on half a mile of water. The fish like rigs. "When them things light up at night, it's like a city. Them lights draw the fish like crazy." Natural is long gone. "It's a lotta fun out there at night," Cody says. We remember fun.

Tommy was born here. Cody, from Tennessee, came for the fishing.

Tommy shrimped for twenty-five years. "Hard work, the open Gulf is. You're up three days straight, takin' heads off shrimp. You're just sittin' there poppin' heads, packin' anywheres from ten to twenty boxes a day." A box is one hundred pounds. How many shrimp?

"Depends on what sizes you're catchin'. You get anywhere from forty-fifties to ten-fifteens. That's how many to make a pound." About the Oil: "I ain't no scientist or anything like that, but I don't think it's gonna be okay in a year. It'll be years, prob'ly. That's just my opinion. It'll be parked a *long* time there."

"That's too scary to think about," Cody says. "I depend on these coming two months for a large portion of my living."

Let's reminisce: "Best snapper fishin' in the world, right off our coast," Cody proudly crows. "Red snapper's our bread and butter. They average five to ten pounds. We also catch vermillion snapper, triggerfish, grouper, amberjack—"

Enough reminiscing. What does he think of the booms and stuff? "A waste of time."

Our boat rocks and growls forward. Our boat is slow and loud. Its noise helps cancel my thoughts. A good thing here, not thinking is. Not thinking about how much I always loved being at sea.

Cody spots a loggerhead turtle that seems fine. It dives. A helicopter passes.

Cody says that a few days ago, there was a slick here that ran for miles. Not here now. It's a stealthy adversary.

Our boat scratches the sea, and on the breeze a whiff of hydrocarbons. Scratch and sniff. Twenty miles into the Gulf we enter a wide mosaic of mazelike slicks freckled with blobs of crude. In the beating heat of the sun, each blob bleeds its own mini-slick into the overall sheen, like pats of butter melting in a skillet.

Under that buttery petroleum coating I am surprised to see small living fishes. I am surprised to see fish alive in the ocean.

A few flyingfish leap through the oil. I wonder how long they can last. Surely they ingest it. Surely their gills are getting gummed. More little tunny chase the small fish in a froth of white explosions that slice the rainbow sheen to ribbons. Life as always, the eternal sea, now with a twist: Can it last, can it stand it? I can't.

Cody says last week, seven miles from shore, he saw oil blobs the size of cars. He was out for opening day of a fifty-seven-day red snapper season. It lasted two days before the government closed the waters.

Heavy lines of crude like ten-foot anacondas, like cobras with hoods spread, for miles now. And snake oil on the breeze.

Cody says, "It was just an accident. You know, we gotta have that oil. It's a necessary evil."

LATE JUNE

In a rather extraordinary breaking of ranks during a congressional hearing on June 15, oil executives distance themselves from BP. "We would not have drilled the well the way they did," says ExxonMobil's CEO. Chevron's chief says, "It certainly appears that not all the standards that we would recommend or that we would employ were in place." Shell's CEO: "It's not a well that we would have drilled in that mechanical setup." In a seemingly sincere apology, BP's chairman says, "This tragic accident should never have happened." Then he manages to offend everyone in the Gulf region by adding, "We care about the small people."

And all across the Gulf, tongues flicker with phrases like "They're no greater than us," "We don't bow down to them," and "We're human beings."

Big People, small people. How sad.

On June 16, a new and improved containment system hooked directly into the blowout preventer begins carrying 5,000 to 10,000 additional barrels per day to another vessel called the *Q4000*, which has no storage capacity and wastefully burns off all that oil and gas. By late June they're either taking or flaring off as much as 25,000 barrels of oil daily. But plenty of oil continues billowing into the sea throughout all this. A third vessel, the *Helix Producer*, is scheduled for hookup to another valve on the blowout preventer by around the end of June. It can burn 25,000 barrels of oil a day. What a waste of everything. What a mess.

Tony Hayward gets a fourteen-page letter from Democratic Congressmen Henry Waxman and Bart Stupak. The letter contains nothing he hasn't already heard and everything he won't affirm. "Time after time, it appears that BP made decisions that increased the risk of a blowout to save the company time or expense," the lawmakers write. "If this is what happened," they venture, "BP's carelessness and complacency have inflicted a heavy toll on the Gulf, its inhabitants, and the workers on the rig."

President Obama visits the Gulf for a fourth time, trying to boost tourism by eating in local restaurants. He acknowledges the tragedy and—responding to public criticism—speaks more harshly about BP.

The White House is also pressuring BP to put tens of billions of dollars into an escrow account. As one of the world's three largest oil companies, BP generates $8 billion to $9 billion every *quarter.* It spends $5 billion to $6 billion a quarter. The difference—$2 to $4 billion—is its average profit every three months. Under the corporation-shielding federal liability cap, BP is legally on the hook for just $75 million—which, for perspective, is 1.25 percent of BP's $6 billion *first quarter* profit this year. If Obama can simply strong-arm billions out of BP, it will be a stunning and masterful coup.

But, needless to say, many Senate Republicans are accusing Democrats and the White House of trying to exploit the oil "politically." They also accuse Democrats of using the calamity in the Gulf to push "a job-killing climate change bill." It's a terrible irony that the blowout has dampened—not whetted—what little appetite there had been in the U.S. Senate to cap greenhouse gas emissions. That's a shame, because we really need an energy bill that puts people to work in new jobs, building the energy future. Without taking from this event a propelling motivation toward new jobs for new energy, the whole blowout becomes simply a calamity, with no lessons learned, no upside, no value added in honor of the lives lost and the lives so changed. South Carolina Republican senator Lindsey Graham had worked on a climate-change bill for months, but has pronounced it hopeless.

How much oil? The company is now funneling about 16,000 barrels a day from its leaky containment cap to a collecting ship. Federally convened experts currently estimate that the well has been spewing 35,000 to 60,000 barrels a day. Every new official estimate is higher than the previous one. Asked how much he thinks is leaking, our Thadmiral demurs: "That's the $100,000 question." He later adds that he believes the figure is closer to the low end of the new estimate, 35,000 barrels. Of course, for weeks he'd acquiesced to 5,000. BP's chief operating officer now says that by the end of June, BP plans to be able to capture more than 50,000 barrels a day.

Meanwhile, oil blobs and slicks coming ashore between eastern Alabama and Pensacola provoke Orange Beach's mayor to rail, "BP isn't giving us what we need. We're screaming for more. We want to skim it before it gets here."

A year after the *Exxon Valdez* ran aground, lawmakers passed the federal Oil Pollution Act to ensure a quick and effective response to oil spills. Every region of the country was required to have a tailored contingency plan.

But the plans amount to what the oil industry says on paper, not a demonstration of what it can do—or what might be needed. The Gulf plan considers a blowout of 240,000 barrels a day into the Gulf for at least one hundred days, far worse than the current leak—yet it declares that "no significant adverse impacts are expected" to beaches, wetlands, or wildlife.

The Minerals Management Service approved that plan. The Thadmiral now says the response has been "adequate to the assumptions in the plans." He adds, "I think you need to go back and question the assumptions."

Thank you, Thadmiral. That's been tried. A 1999 Coast Guard report recommended that equipment like booms, skimmers, and absorbent materials should be increased by 25 percent. But over the next several years, lobbyists for oil companies pushed to keep the existing standards in place—and emphasized the cheaper alternative: chemical dispersants.

In August 2009, the Coast Guard effectively overruled its 1999 report, declining to require the substantial increase in the amount of mechanical response equipment.

"BP could fire all their contractors because they're doing absolutely nothing but destroying our marsh," rages Billy Nungesser, the president of Plaquemines Parish. Weeks in, he says, "I still don't know who's in charge; is it BP? Is it the Coast Guard?" "The boom has been a disaster from the beginning," Florida senator Bill Nelson tells a Senate hearing. "You have a big mess, with no command and control." Florida's attorney general says he's "absolutely appalled." The mayor of Orange Beach, Alabama, says it's "a very discombobulated and discoordinated effort."

By now people have found more than 350 dead or moribund loggerhead turtles along the Gulf Coast since the blowout began—including 20 carcasses in a single day. More than 60 have been covered in oil. Fishing gear remains suspect for some of the deaths. Meanwhile, more than 70 human Louisiana residents have reported oil-related illness to the state's Department of Health and Hospitals.

Morning. Another day of this. I've driven from the Florida border, across Alabama and Mississippi, and into Louisiana, and I'm watching dawn rise in the rearview mirror. What an awful night. In the last couple of weeks our dog, Kenzie, had lost her appetite. Yet just a couple of days ago she was walking with her tail high, giving us hope that she'd get better. After I left for the Gulf she weakened rapidly, and last night she was suddenly unable to walk. Patricia called me at midnight and took her to an all-night veterinary clinic. They found a tumor in her spleen and blood in her body cavity. The vet said that operating would not guarantee success even in the short term, and partly because of Kenzie's age, approximately thirteen, he recommended euthanasia. Pat sat with Kenzie for a while, then called me tearfully at 2:00 a.m. to say our dog was gone. I will always remember our walks on the beach and watching her run so far along the ocean, or alongside us as we biked around Lazy Point; and, more recently, helping us herd our new chickens back into their coop, and being so interested yet gentle with the infant raccoon that fell to earth in our yard. I regret that I was not home for this crisis. I regret that Patricia had to shoulder the burden alone. I regret being here. Busyness hurts relationships.

Dew and low fog hang heavy on the grass. Eighty-three degrees by 6:30. Roads lined with pines and worry.

Radio news: Louisiana's Governor Bobby Jindal is accusing BP of dragging its feet on paying claims; President Obama will meet today with grieving relatives of workers killed in the explosion. The president wants to include deepwater drilling "as part of a comprehensive energy strategy." The radio says wildlife workers are picking up birds on islands. There's concern about oiled pelican nests.

I think they should destroy all the nests and try to frighten the adults away; that would break the adults' motivation to keep returning and continuing their exposure. For the population's future, saving breeders is more important than trying to save eggs or even chicks. If adults die or get immobilized and rescued, their eggs and chicks are doomed anyway. That's uncomfortable to say, but it's true.

Along the interstate, signs for a concert venue advertise coming headline acts: Liza Minnelli (still?), Ringo Starr *(still)*, the O'Jays (really?). The highway reaps its constant harvest of armadillos; our most heavily armored surviving mammal cannot survive our daily onslaught. More to pity. An alligator with an urge to roam has had its ambitions crushed. Glad I wasn't in that car.

The sign reminds travelers, "Welcome to Louisiana—America's Wetland." Almost immediately the land seems lusher, greener, lower, wider. The rest area is stocked with pamphlets for tourists: fishing, seafood. Oh, well.

Radio talk show: A guy who runs a parasailing business calls to say, "Congratulations, BP. Today you've accomplished something hurricanes couldn't do, banks couldn't do, and even the greed of Wall Street couldn't do: you've put me out of business." He has made more than eighty phone calls trying to get BP to respond to his financial claims.

Three congressional hearings today. Five yesterday.

Kevin Costner has a solution to the oil; for years he's invested in a machine to separate oil from water. There are thousands of other people making suggestions, promoting their ideas, hawking their own products. There seems no clear route for these things to get evaluated.

Time to have another look as the frigatebird sees it. Wheels defy gravity at Belle Chasse, Louisiana, a little after 10:00 a.m.

Our twenty-four-year-old pilot is Corey Miller. Up over the marshes, we're oil hunters once more. Soon we're flying in and out of clouds, and this makes me nervous. Increased air traffic in the last few days has brought planes close enough to blow kisses.

The curious and the dispersant dispensers and the helicopters crowd the sky. Slip-sliding by. Up here at these speeds, gaps close fast. You want good vis. We have bad vis.

Sometimes we're whited out for minutes. Cat-and-mouse with other planes. Corey talks to his headset constantly. Young and alert, I hope; not cocky. Seems on the ball.

A rather alarming number of wildlife biologists get raked off in small-plane crashes. Do a computer search with the words "biologist plane crash" and you'll see what's on my mind right now. Once, in Honduras, I was about to board a small plane to an outer island, but agreed to go later so some other people could get back to the mainland first. During their flight the engine quit, and the plane crashed into the sea and sank. Luck put them close enough to land that someone saw the plane plummet and raced to pick up survivors. But I'd planned a much longer water crossing, and if we'd done my trip first, we'd probably never have been found.

Yet right now, I'm not worried about our plane. I'm worried about other planes. Our current destination, directly above the blowout, is also of much interest to the other fliers. Corey tells me they've worked out altitude separations for the different aircraft. I tell him good, thanks. I gaze down, and think of home.

Headed to 3,500 feet, for safety's sake. A little high for seeing details well, but you see more if you're alive. The clouds let us play peekaboo with the ocean. They also cast Rorschach-test shadows. We see: barges, boats. Oil rigs, of course. Muddy troughs of river water slinking seaward, soils of the Great Plains, carrying their fertilizers and pesticides, maintaining the Gulf's chronic illness. Like I said: problems besides petroleum.

The shimmering sea pea green now. The engine noise. When we open a side window for photos, the flapping wind.

Shrimp boats continue towing boom through thick and thin oil with no significant effect on it. Most oil spills over their U of boom. Slicks greet the boats' bows, slicks are their wake. They've been wasting time like this for weeks now.

We apply more altitude to defeat the thickening clouds. A worse and safer view of a sea streaked widely with oil. The sea keeps appearing and disappearing, blue sky appears and disappears. The one constant: oil.

Miles and miles of streaks, tendrils, fingers. Oil coats the ocean brick orange. Brown. Differing densities under varying light. Miles, miles, miles.

With his small camera, Corey is taking pictures. I remember being twenty-four. So will he.

At 11:30 we're flying ellipses over ground zero. The clouds break nicely, revealing a couple of *dozen* ships. Ships large and smaller. The relief well drillers. The flare of burning gas. Choppers touch ships' helipads like dragonflies grasp reeds.

The blue water looks like mere cracks between heavy brown billows of used motor oil. An absolute mess. The horizon-gobbling *scale* of this now.

My camera loads image upon image to its buffer before I give it a chance to save. Maybe I can get it all, take it all out of the ocean.

Corey's had my headset turned off while he's been talking to planes and air-traffic control. He hits a button and says to me, "A photo makes it look like one area. Until you get up here and see that it's as far as you can see in all directions. You can't get that in a picture."

Picture that.

At 12:15 P.M. more or less, we're back over the shoreline. Thick blobs near the shore. Waves lap darkly. The beach is stained. The emerald marsh, bordered black.

But only bordered. I sense the marshes' restorative power. Nature hurls hurricanes, droughts followed by floods, diseases, famine. But in the long haul—and this will require time—nature is what can get us out of the trouble we keep getting ourselves into. They call it Mother because, though she can punish, she's why we're here in the first place. She's the hope we have while we're so hopelessly juvenile.

As the heat of June soars, anger mounts over the Obama administration's recently declared six-month moratorium on exploratory deepwater drilling. The president is in a no-win spot, with people both demanding that he do more and condemning him for what he is doing. The goal was to give the government time to review the rules for oversight of such wells. Many Americans horrified by the blowout welcomed the shutdown.

But here, the moratorium on new drilling is deeply unpopular. Various people are calling it things like "a death blow to Louisiana" that "has and will continue to destroy tens of thousands of lives," creating "an economic ripple effect that will be catastrophic to our entire region."

Those working in the petroleum industry see the moratorium as a bigger threat than the spewing oil. Many of the affected rigs will seek to drill in other countries, imperiling an estimated 800 to 1,400 jobs per rig, including third-party support personnel. As many as 50,000 jobs may be affected, though the industry and Louisiana's elected officials provide the highest estimates.

"From 1947 until 2009, there were 42,000 wells drilled in state and federal waters in the Gulf of Mexico and 99 days ago one of them blew up," rages Democratic senator Mary Landrieu of Louisiana. "But no matter how horrible that is, you don't shut down the entire industry." She says that a temporary ban on drilling, even if it lasted for only a few months, could affect as many as 330,000 people in just Louisiana.

I'm not downplaying the dilemma here, but I don't trust her numbers or her hyperbole. The Louisiana Economic Development department estimates that the six-month moratorium will cause the loss of up to 10,000 jobs within a few months. State figures show that the area's whole oil and gas industry—not just the deep exploratory drilling that is temporarily banned—supports 320,000 jobs, $12.7 billion in wages, and $70.2 billion in business sales.

To the industry, its supporters, and its congressional backers and lobbyists, a moratorium doesn't make sense when more than 30,000 other wells have been drilled in U.S. waters, 700 of which are as deep as or deeper than the one the Deepwater Horizon was drilling. And they argue that the incentives are already there to avoid the fate of this ill-fated well; no one wants the kind of mess BP now has all over its hands. One rig worker says, "For us to stop drilling in the Gulf is like ending our lives as far as the way we live. It's really that scary." That's what fishermen say about their inability to keep fishing. (The administration will lift the drilling moratorium on October 12.)

President Obama signals what might be the first inflection point from unmitigated disaster to silver linings. It's part rhetoric, part counterbalance to the panic. He promises, "Things are going to return to normal." Not only that, but: "I am confident that we're going to be able to leave the Gulf Coast in better shape than it was before." He declares seafood from the Gulf's open fishing areas safe to eat. And he announces that he's pried from BP an agreement to set up a compensation fund that will run into "the billions of dollars."

In the same span of days, the biggest mess to come ashore so far is heavily coating Louisiana's enormous Barataria Bay. Birds sodden in syrupy crude, stranded sea turtles, beaches blanketed in brown goo, marshes bathtub-ringed in oil produce some of the most heart-wrenching photos yet of the blowout.

"This was some of the best fishing in the whole region, and the oil's coming in just wave after wave. It's hard to stomach, it really is," says one fishing guide. A resident adds, "We got little otter families that swim in and out, we got coons, all that good stuff, man. It's good for the kids out here. They swim, work on the boats, fish." They did.

Day and night, the well continues injecting more of the same into the oily Gulf. It feels like a siege. Like it's hopeless.

Because I'd wanted to fly right over the blowout, we had to stay high. But in a different part of the Gulf, author David Helvarg and conservationist John Wathen were low enough to see wildlife. Their video is the most affecting thing I've seen yet about the blowout's ongoing effects. In a TV interview on MSNBC, Wathen, in a mellow Mississippi drawl, describes seeing dolphins mired in oil. "You can see the sheen for miles and miles and miles to the horizon. We figure we saw over a hundred dolphins that were in distress. Some were obviously dead, belly-up in the water. And others, they looked like they were in their death throes." He talks about seeing the Gulf set aflame, the towering columns of smoke.

Still ninety-three degrees at 5:00 p.m. Everything here is far from everything else. Grand Isle, on the seacoast side of Barataria Bay, is about a two-and-a-half-hour drive from New Orleans. The road runs along canals for many, many miles, the roadside vegetation beautiful, lush, green, summery. I especially enjoy seeing the draping Spanish moss hanging from trees. And those black vultures in that robin's-egg-blue sky, effortlessly circling.

Lots of modest to run-down homes. Not a thriving place. Looks like money is tight. Beauty salons in people's homes; their signs on the lawn. A sign advertises frogs' legs and turtle meat. "Ducks for Sale." Live minnows, live crabs. Docks every now and then. Boats. *Bayou Queen*, *Lady Catherine*, *Captain Toby*, looking both proud and forlorn, tied up, underemployed. "For Sale: Black Angus Bulls and Heifers." Mitch's Garden Center. Doc's Body Shop. The Flower Pot florist. Debra's Movie World. Tiresome billboards picturing real estate agents, insurance agents, car salesmen. A funeral home's hand-painted sign. Austin's Fresh Seafood bears a smaller hand-painted placard: "Closed Due to BP."

Up over the Intracoastal Canal.

And now displays of grief and rage come bubbling to the surface. At one intersection, murals. Obama's portrait and the words "What Now?" A Grim Reaper identified as BP, captioned, "You Killed Our Gulf. You Killed Our Way of Life." A grim statue of a person holding a dead fish, accompanied by a child whose head is bowed, painted as though oil-drenched, labeled, "God Help Us All."

On the main road, with a speed limit of fifty, I suddenly notice I'm doing eighty. It's not that I'm in a hurry to get there. It's that I'm in a hurry to get this over with so I can leave. I'd better watch my speed; a cop is occupied with someone who didn't watch his.

The road narrows to one lane. The paralleling channel widens as I near the coast. Thirty miles to go. There are some pretty big ships here now. More port facilities for oil rig tenders and the like. More industrial plants related to oil and gas. Not quaint.

At the bridge to Grand Isle, the horizon is punctured by derricks, giant antenna towers guy-wired into the marsh, petrochemical tanks, helicopter pads, warehouses on stilts, tugboats and rig tenders. Chevron's aircraft operations has its own small airport for helicopters.

While I'm driving through Leeville, the radio conveys that "BP said it was unaware of any reason for the stock price drop."

I'm on a road paved right through an immense marsh that stretches from horizon to horizon. How many millions of wildfowl must once have swarmed into here. I've stopped in three places looking for a road map. No one sells them. But in each place stood a line of guys buying beer. I guess that's what there is to do.

Contrast: In places like Shell Beach and Hopedale, the fisherfolk seem woven into the place like vines grown up on netting. Can't be uprooted. Transplantation would be wrenching, possibly impossible. The rig workers, different. Just a job. Could be anywhere. Might as well be. Looking displaced. Their workplace distant and forsaken from whatever counts to them as home. Painful to witness. Everything here in pain, or bored.

It's hard to imagine that anything but oil could have made it worth the time and money to build a raised roadbed and pave it. This road seems so tenuous, so vulnerable to harsh weather. All the houses here are on stilts. A Forster's tern dives into a creek. Surprisingly, pleasantly, there are plenty of white, un-oil-stained egrets here.

And suddenly a cheerful "Welcome to Grand Isle." Stylish roadside sign, letters three feet high. Blue stylized waves harken back to when the ocean was that color. Artful and colorful, of cut steel, with colorful steel marlin and redfish and tarpon swimming across it. Flanked by planted palms. Pretty and proud.

Anxious for a quick glance at the water, I turn where it says, "Welcome to Elmer's Island. Open Daily." When a sign warns, "Access at Your Own Risk," I take it at its word. Where the sign says, "Beach Closed," I continue.

Before reaching the telltale lineup of Porta Potties that have become BP's major expression of Gulf architecture, I get repelled by security guards incredulous that police did not intercept me sooner. (Who'd have thought that Portosans, so innocuous at Woodstock, so reliable during the Age of Aquarius, would turn ominous and foreboding so soon into this millennium?)

"If you come back here, you'd have to get a BP representative to come with ya. This is a BP safety area. You need a hard hat, steel-toed shoes, safety glasses—"

Another impressive display of BP's near obsession with keeping everyone *safe.* In the wrong ways, at the wrong time, sweating the meaningless small stuff.

"They got containers full of oil. You get a whiff of that crap—I don't know; they just don't want nobody passed out, y'know?"

No, I don't. Because hard hats, steel shoes, and safety glasses won't protect you from fumes. This is the company that refuses respirators. Talk to anyone with an ID badge, you can't help feeling every word is bullshit. It's not always their fault. Sometimes they're just regurgitating the bullshit they've been overfed. Some of the bullshitters are nice: "If you get one of them BP guys to come with ya, you can come back," he offers as I'm turning around. And some, less so.

This is a miserable purgatory of a place. The channels cut straight in a place where water naturally meanders. The raised roads trespass into low country where boats would belong. And now there's oil where water belongs. Oil where honesty belongs. Everything at odds with the place's soul.

It's a contest for the worst kind of possession, the one that diminishes what it acquires, harms what it strives to hold.

Grand Isle is grand indeed, many miles long. Marshes and power lines on the north, sandy Gulf beach on the south. A horizon pierced and pincushioned by cranes. A weathered sign says, "Jesus Christ Reigns over Grand Isle," but another new "Beach Closed" warning indicates that BP reigns now. I pass sheriffs and a couple hundred day laborers climbing aboard school buses in a big lot full of Porta Potties.

I duck down a road. On the beach: Porta Potties. Little shade shelters spaced evenly along miles of shoreline. Stacked cases of water. It's already after quitting time, 7:00 P.M., so no one's here. It's still hot, and a bottle of water would be nice. Perhaps BP owes me a little *clean* water. I forbear. I don't actually want anything of BP's. As far up and down the beach as I can see the dry sand is thoroughly crisscrossed with tire tracks. Was a beach, is a disaster site. BP's D-Day.

There's no one here to watch the sunset, stroll hand in hand, look for shells, or take an evening dip as the day releases its hot irons of heat. One woman leads her small son along the sand. Having a hard time negotiating the corrugated ridges and dips of tire tracks in soft substrate but trying to make a go of it, he follows like a toddling bear cub. She tells me she can last a year like this; after that they don't know. Says we can be here, but can't go past the berm to the water. It's patrolled. I see one distant vehicle moving on the sand. When she leaves, I walk over the berm anyway; I need a closer look. Nearer the water the *entire* beach is oiled. Long, dark band of stain. Fresh blobs and splatters.

Rather surprisingly, the water seems like water. Where the waves lap, there's some clarity to it. And I can see some crabs, alive. The Oil comes and goes in great waves. Unpredictable foe.

A few days ago, oil slurped the shore so thickly here that pelicans looked cast in bronze. Horrible. That massive murk has moved. The water from the shore outward remains slicked and splatter-dappled, but it's not a black lapping mat right now.

Two middle-aged women get out of their car. Just drove sixty miles. "I never thought I'd see anything like *this!* It used to be a sand beach. You could come out and have parties and picnics and swim. This is really—"

"You think *God* would do something about this."

"The hurricane season, it could pick up all that oil out there and put it all over everything."

A small band of souls come out to set up their volleyball net in the tire tracks. Resolute, jaws set for fun.

Outside a cottage, a sign: "BP Headquarters." Its arrow points directly down into an actual toilet bowl.

Closer to the east end of the island and on the north side, mullet jump. Gulls loaf. They look okay. Terns fishing along the bridge offer cheer that even here life continues. Overhead, those pelicans are all soiled. But still flying, at least. At least for now.

Speckled Trout Lane. Bayside Circle. Redfish Lane. Pete's Wharf Lane. Sunset Lane. A place for sun and fun. Was. News networks, their trucks parked at cottages, provide some owners with summer rental income. At other cottages, camouflaged fat-wheeled trucks. Varied contractors and the National Guard.

Overhead, a frigatebird. *Really* brown pelicans. Distinctly dingy, they skim over the slick surface. I fear for them.

On a lawn, a graveyard of white crosses memorializes these departed: "Beach Sunsets," "Sand Between My Toes," "Marlin," "Sand Castles," "Dolphins," "Bluefin Tuna," "Crabbing," "Shrimp," "Sailing," "Beach Sunrises," "Summer Fun," "Sea Turtles," "Picnics on the Beach," "Floundering," "Flying a Kite," "Sand Dollars," "Oysters on the Half Shell," "Boogie-Boarding"; there're about four dozen more.

Nearby, a much larger cross says, "In Memory of All That Was Lost; Courtesy of BP and Our Federal Government." Another cross marks the passing of "Our Soul." Another roadside sign: "BP—Cannot Fish or Swim. How the Hell Are We Suppose to Feed Our Kids Now?" Signed by the owner. A hurting, hurting place.

Based on the latest flow rate estimates of up to 60,000 barrels per day, the fine for the escaping oil alone could be $260 million *per day.* Anyone still doubt that BP has been trying to hide the body? Criminal penalties, if fully imposed, could cause the costs to balloon to more than $60 billion, dwarfing an escrow account the White House wants BP to establish for paying claims of economic loss.

The government will likely use BP's prior criminal record, such as BP's guilty plea in the 2005 refinery explosion that killed fifteen people in Texas City, to argue that the Deepwater Horizon disaster resulted from a corporate culture that lets hurrying and cost-consciousness jeopardize safety.

BP's carefully crafted public image of friendliness belied its egregious record for serious safety problems. The *Economist* reports that between June 2007 and February 2010, BP received an astonishing 97 percent of all operational safety and health citations for "willful" and "egregiously willful" breaches of the rules at American oil refineries, adding that this is "a remarkable share even allowing for close scrutiny after Texas City."

But the laws that would be brought to bear generally don't have felony provisions that would lead to jail time for executives, and where they do, prosecutors would have to directly connect a defendant with a crime.

On Capitol Hill, congressional Democrats Henry Waxman and Ed Markey blast the heads of ExxonMobil, Chevron, ConocoPhillips, BP, and Shell Oil for producing "virtually identical" disaster response plans. All discuss how to protect those famous walruses in the Gulf of Mexico. The congressmen excoriate the oil titans' "cookie-cutter plans," citing sections that have "the exact same words," indicating an investment of "zero time and money."

It will turn out that five giant oil companies all got their response plans from the same tiny Texas contractor. The firms all assured the government that they could handle oil spills much *larger* than the one now threatening the region's environment and economy. And each time, the Minerals Management Service approved the plan and gave the go-ahead for drilling.

In its exploration plans for Alaska, Shell has analyzed the prospect of only a 2,000-gallon diesel fuel spill. It asserted that a larger crude oil spill would be unlikely because the water is shallow. The Minerals Management Service skimmed up this assumption without questioning it. Shell is relying on a single company based about three hundred miles from its intended Chukchi Sea drilling sites. Anything goes wrong there, Shell would have available only a tiny fraction of the resources BP called up in the Gulf.

Big ol' jet airliner. I doze, then rouse. That's Sandy Hook, New Jersey. I glance at the ocean and reflexively look for streaks of oil. The stain in my brain.

On Long Island, when an egret flies over my house, I absentmindedly check it for signs of oil, as has become my habit. At our local marina, they're talking the Oil. At the beach, it's sun lotion and the Oil. It sticks to everyone's minds. Follows everywhere. You can't wash your thoughts of it.

At the marina's outdoor bar, the loud ones blame environmentalists. (That's what their favorite broadcasters have told them to say.) "There's no reason not to get the oil that's in Alaska." "Because they won't let them." "It's the enviros." "The enviros pushed them to this; BP was pushed into this by the enviros." "Big guvvamint." "Can't do nuthin' anymore."

In his first and much-anticipated address to our nation from the Oval Office, President Obama called for a new "national mission" to wean the United States off fossil fuels. "The tragedy unfolding on our coast is the most painful and powerful reminder yet that the time to embrace a clean energy future is now. Now is the moment for this generation to embark on a national mission to unleash American innovation and seize control of our own destiny."

Perfect!

But why isn't Obama throwing all his weight behind the new energy bill unveiled by Senators John Kerry and Joe Lieberman? Because the Kerry-Lieberman bill includes fees on carbon emissions, and the White House is afraid Republicans will ride the slogan "carbon tax" to multiple victories at the midterm elections.

It's hard to see how America can accomplish anything as long as two parties locked in a death battle can't see past two-year congressional cycles. But I think Obama should press it, because he will never win his opponents over, but by not acting boldly he is losing the enthusiasm of his supporters.

Thomas Friedman had written that this is not Obama's Hurricane Katrina; it's his 9/11—one of those rare seismic opportunities to energize the country to do something really important that is too hard to do in normal times. But as Bush blew 9/11's possibilities, Obama is blowing this blowout. To Americans wanting to do something for the country they love, Bush told a few to go fight and the rest of us to go shopping.

Boosting new energy technology and building new energy infrastructure would seem to line up with the mood among those who elected Obama to bring sweeping reforms. Yet Obama is chary, and a well-oiled moment is slipping through his fingers.

Obama, having already asked congressional Democrats to make a hard vote on health care, seems to feel he can't ask them for another. He isn't publicly deploying his assembled brain trust, including Nobel Prize–winning Energy secretary Steven Chu, to rally Americans who are waiting to be enlisted, ready to be rallied.

When people ask me, "What can I do to help the Gulf?" I don't know what to tell them. There are no real opportunities for the public to just go down and help out, and for the Big Picture, we haven't been given a concrete presidential vision that we can get behind.

"Mr. President," Friedman offers, "Americans are craving your leadership on this issue. Are you going to channel their good will into something that strengthens our country?—'The Obama End to Oil Addiction Act'—or are you going to squander your 9/11, too?"

In a similar but much more enjoyable vein, MSNBC's Rachel Maddow assumes the role of fake president just long enough to deliver the speech she wishes the president had given. Among other things, she wishes he'd said:

> Never again will any company be allowed to drill in a location where they are incapable of dealing with the potential consequences. . . . I'm announcing a new federal command specifically for containment

> and cleanup of oil that has already entered the Gulf of Mexico. . . . I no longer say that we must get off oil like every president before me has said. We will get off oil and here's how: The United States Senate will pass an energy bill. This year.
>
> When the benefits of drilling accrue to a private company, but the risks of that drilling accrue to we the American people, whose waters and shoreline are savaged when things go wrong, I as Fake President stand on the side of the American people, and say to the industry: From this day forward, if you cannot handle the risk, you no longer will take chances with our fate to reap your rewards.

Maybe Maddow will someday throw her sombrero into the presidential ring. And if she wins, she will join every president since the 1970s in saying that America must get off oil. Richard Nixon, Gerald Ford, Jimmy Carter, Ronald Reagan, George H. W. Bush, Bill Clinton, George W. Bush, Barack Obama. And Rachel Maddow.

But until the country realizes that our Congress and our courts must serve people with belly buttons, not multinational corporations . . .

More congressional hearings and briefings. The Interior Department's acting inspector general, Mary L. Kendall, tells a congressional panel that in the Gulf region the Minerals Management Service has only sixty inspectors to oversee about four thousand drilling facilities. Inspectors in the Gulf operate "with little direction as to what must be inspected, or how." Yet on the Pacific Coast, ten inspectors cover only twenty-three facilities. She says the minerals service has a difficult time recruiting inspectors because the oil industry tends to pay a lot more.

The sargassum weed, whose yellow floating mats provide cover and nursery habitat for many kinds of sea life in the open Gulf, is dying. Most people have never seen or even heard of sargassum, but it shelters and feeds uncountable numbers of fish and young sea turtles. Tunas, mahimahi, billfish, mackerels, and others often haunt its edges. Blair Witherington, a research scientist and sea turtle expert with the Florida Fish and Wildlife Conservation Commission, says, "Ordinarily, the sargassum is a nice, golden color. You shake it, and all kinds of life comes out: shrimp, crabs, worms, sea slugs. It is really just bursting with life. It's the base of the food chain. And these areas we're seeing here by comparison are quite dead." He speaks of seeing flyingfish land on rafts of oil and get stuck right there. He says the jellies are dying. "These animals drift into the oil lines and it's like flies on fly paper," Witherington relates. "As far as I can tell, that whole fauna is just completely wiped out." Of dispersants, he says the thinking is "just keep the oil out at sea; the harm will be minimal. And I disagree with that completely." Of his beloved turtles, he says, "We've seen the oil covering the turtles so thick they could barely move, could hardly lift their heads."

It's hard to imagine a turtle in the ocean catching fire, but after all, it's hard to imagine any of this. And so today's vocabulary word from the theater of the absurd is "burn box." Noun: an area of corralled oil set on fire. Contractors are staging mass burns of some of the floating oil mats, setting the sea aflame. Some people allege they're burning up sea turtles along with the oil. I'm dubious. But then again, a lot of little turtles could be clinging to weed mats. The turtle rescuers want to pick up as many turtles as they can find, for fear they'll be incinerated. They find eleven, all of them heavily speckled with oil.

I read that only 3 percent of the slick is thick enough to catch and hold a flame on the surface of the sea. But there is so much oil that the fires and their towering billows of thick smoke are horrendous.

Workers will light more than three hundred fires at sea, sending thick plumes of smoke, carbon dioxide, and hydrocarbons toward heaven. They'll burn more oil than the *Exxon Valdez* spilled. A man employed to ignite floating oil will brag, "No one can deny this is a success."

John Wathen, who spoke of seeing dolphins dying in thick oil, says he witnessed other dolphins, lined up with their heads out of the water, watching the astonishing sight of their ocean in flames.

Various people report seeing sharks, mullet, crabs, rays, and small fish in unusual numbers close to shore. Are they fleeing the oil? Could be. But the Gulf has long had its dead zone of low oxygen, which worsens in summer. Large numbers of fish moving into shallows where there's more oxygen isn't unheard of. The latest figures of dead wildlife total about 800 birds, 350 turtles, and 40 mammals. It's not clear whether the oil killed them all. Even people cruising Louisiana's heavily besmirched Barataria Bay see dozens of dolphins frolicking in oil-sheened water, and oil-smudged pelicans feeding their young.

In some areas still open to fishing, fishers report large catches of red snapper, grouper, king mackerel, and amberjack. Are the fish congregating after fleeing the oil or are they unaffected by it? Are the large schools of fish locals see hanging around piers there because the fishing ban has given the fish a huge break? No one knows. But it's seeming that much life in the Gulf has resilience enough to resist the oil.

After many days of very public pressure from President Obama and many hours of private negotiations, BP finally agrees to divert $20 billion into an independent fund to pay claims arising from the blowout. The company will also suspend paying shareholder dividends for the rest of the year and will set aside an additional $100 million as compensation for lost wages for oil rig workers affected by the Obama moratorium on deep exploratory drilling. The president stresses that the amount is not a ceiling on BP's obligations. Suspending dividends delivers another blow to BP's reputation and its shareholders, but for the company's accountants it's a godsend that saves BP something like $10 billion. The president had earlier alluded to his determination to step in and do "what individuals couldn't do and corporations wouldn't do." For the president and the Gulf, it's a stunning coup.

To Obama, this is a rebalancing after two decades in which multinationals sometimes acted like mini-states beyond government reach, while influencing the government to, as he says, "gut regulations and put industry insiders in charge of industry oversight." The president had no legal basis for the demand. (Remember, BP is legally on the hook for just $75 million.) The deal follows an extraordinary four-hour White House meeting that was punctuated by breaks as each side huddled privately. BP will pay into the fund over four years, at $5 billion a year. Last year, BP generated profits of $17 billion. After the announcement, BP shares close up 1.4 percent, at $31.85.

While the president may have been walking a fine line, at least one member of Congress was blocking the intersection. Republican congressman Joe Barton of the great oiligarchy of Texas rails that Obama acted illegally, and—during a congressional hearing—Barton apologizes to BP executives for our president's "shakedown" of their company. That rumbling is the sound of jaws dropping across America. Even though Barton had reportedly gotten $1.5 million in campaign donations from oil companies, his outburst is bizarre enough to quote at length: "I am ashamed of what happened in the White House yesterday, that a private corporation would be subject to what I would characterize as a shakedown," said Barton. The fund, he said, "amounts to a $20 billion slush fund that is unprecedented in our nation's history" and "sets a terrible precedent for the future." He continued, "I apologize . . . I do not want to live in a country where any time a citizen or a corporation does something that is legitimately wrong it is subject to political pressure that amounts to a shakedown." Keep in mind, this guy was in charge of the Energy and Commerce Committee before the Democrats won the House majority in 2006. And as the political pendulum swings, he'll likely be the chairman again. One thing we agree on: I don't want him to live in this country, either.

On June 17, a *New York Times* editorial opines that BP's CEO, Tony Hayward, has just given Congress "a mind-bogglingly vapid performance." Congressmen Waxman, Stupak, and others spent hours trying to pry answers out of Hayward about what went wrong. Mostly, he deflected and sidestepped the grilling and frustrated the congressmen and the American public. "I was not part of that decision-making process" was his frequent answer to questions. But fair enough; that's true. To Texas Republican congressman Michael Burgess, who was taken aback by the idea that Hayward had no prior knowledge of this well, Hayward answered, "With respect, sir, we drill hundreds of wells a year around the world." To which Burgess shot back: "That's what's scaring me now." During the seven-hour hearing, something like 735,000 gallons of oil leaked into the Gulf.

Rather to its credit, I grudgingly admit, BP releases $25 million of a pledged half billion dollars over ten years to support several universities' research into the effects of the blowout. To make recommendations on which institutions will receive funds, BP appoints an expert panel chaired by environmental microbiologist Rita Colwell, who formerly headed the National Science Foundation and is now a distinguished professor at Johns Hopkins University. Sounding so refreshingly out of character that the cynic in me has trouble figuring out BP's motivation, BP CEO Tony Hayward says in a press release, "It is vitally important that research start immediately into the oil and dispersant's impact, and that the findings are shared fully and openly. We support the independence of these institutions and projects, and hope that the funding will have a significant positive effect on scientists' understanding of the impact of the spill."

So, summing: BP has agreed to pay $20 billion, pledged $500 million for a ten-year research program to study the blowout's lasting effects, agreed to contribute $100 million to support rig workers idled by the administration's deepwater drilling moratorium, and paid over $50 million to promote Gulf tourism.

And yet, doesn't it always seem that no good deed goes unpunished? The House is lining up to pass a drilling overhaul bill that, inter alia, would bar any company from drilling on the outer continental shelf if: more than ten fatalities had occurred at its offshore or onshore facilities or if, in the last seven years, it's paid fines of $10 million or more under the Clean Air or Clean Water Acts.

BP is the only company that currently meets that description.

Coincidence? "The risk of having a dangerous company like BP develop new resources in the Gulf is too great," said Daniel Weiss, Representative George Miller's chief of staff. "Year after year after year, no matter how many incidents they're involved in, no matter how many fines they've had to pay, they never changed their behavior. BP has no one to blame but themselves."

BP's bargaining chip: it says that if it can't drill, then maybe it won't be able to pay. Our bargaining chip: in about two weeks the House will, in fact, pass that company-banning language, helping guarantee BP's continued attention.

This, I think, is true: BP, which gets more than 10 percent of its global production out of the Gulf of Mexico, needs us more than we need it. There are other companies that would send the same oil ashore.

Seeming to recognize that fact, BP is—for once—on its best behavior. Not only is its $20 billion escrow agreement with the White House voluntary, but "We have committed to do a number of things that are not part of the formal agreement with the White House," notes a BP spokesman, in case America really hadn't noticed. "We are not making a direct statement about anything we are committed to do. We are just expressing frustration that our commitments of good will have at least in some quarters been met with this kind of response."

I receive this inane e-mail:

> To help clear the toxins from the water, we will be using the energies of love and appreciation, and a special prayer related by Dr. Masaru Emoto. Dr. Emoto is the Japanese scientist who has done extensive research on how the energy of love and appreciation can change the molecular structure of water at the quantum level. The Process: Stand near, or in, the Gulf of Mexico. Or, imagine that you are standing

> there. Direct your thoughts and energies to feelings of love and appreciation. When you are filled with loving thoughts, speak the following prayer: "To whales, dolphins, manatees, pelicans, seagulls, and all aquatic bird species, fishes, shellfish, planktons, corals, algae, and all ion creatures in the Gulf of Mexico, I am sorry. Please forgive me. Thank you. I love you." Join us in prayer. We are one, and the One will join us together in this great work from wherever we are!

So there you have it.

Back on planet Earth, the total count of sea turtles recovered dead, injured, and oiled is up to about 460. The number of turtles coming up to nest has gone down, but the cold winter could have translated into a late spring for them.

Diane Sawyer, ABC News, has this report:

"Who is in charge of the cleanup? For four days we have been asking that question and we have not been able to get an answer. David Muir was with two frustrated governors today. David?"

"This barge should be out in the Gulf sucking up oil. But sixteen of these barges are docked here, all under orders not to move."

Louisiana Governor Jindal: "The Coast Guard stopped them from going to work."

David Muir: "And then today, word from the Coast Guard saying, 'Go ahead.' This kind of confusion is everywhere. Who's in charge here?"

Alabama's Governor Bob Riley: "Great question."

Diane Sawyer: "Two governors saying they cannot get straight answers."

But alongside the video on the Web, the print version of the story says, "The Coast Guard needed to confirm that the boats were equipped with fire extinguishers and life vests." Well, I've been critical of the Coast Guard, but let's be fair: that *is* a straight answer.

On June 20, Tony Hayward "steps down" from being BP's gusher usher (after he was stepped *on* for making so many trips over his own feet). One day later, he has his life back. Off the Isle of Wight he attends yachting races with his son, while BP PR races to defend his right to do so. "No matter where he is, he is always in touch with what is happening within BP," the BP spokesman says of Mr. Hayward, the very man who'd told the U.S. Congress that he was not aware of the Macondo well as Deepwater Horizon was drilling it because "with respect, sir, we drill hundreds of wells a year around the world."

The events prompted Senate Republican Minority Leader Mitch McConnell to say, "All of these guys could use a better PR adviser."

Yes, Tony Hayward has his life back, but Gulf people are saying things like "I see my life ruined. There ain't no shrimping, there ain't no crabbing, there ain't no oystering. Well, the only thing I know is shrimping. That's all I know. Now, you tell me: Where do I go from here?" The owner of a seafood company that normally ships fifteen million pounds a year gets up in the morning, walks to his empty warehouse, trudges back again, sits down in front of the TV, and stares at CNN's oil spill coverage; then he heads back to the warehouse. "I'm just walking around in a circle," he says. "I never been this confused in my life." In the U.S. House of Representatives, while speaking to the House Energy and Commerce Committee, a Lousiana congressman breaks down in tears. He's not alone. A fisherman in his fifties explains, "We start talking, and before you know it, we're all crying. Tough men, you know? Tough as they come. Just break down and cry."

Fisherman: "The first thing I'd like to do is punch that CEO in the mouth. That'd make me feel a little bit better, I guess."

Social worker: "There's breaking points for people. You look at some of these people and you wonder, when is that person going to snap?"

People snap differently. In Alabama, at least one fisherman, despondent, chooses to make his final exit while sitting aboard his beloved boat.

And as one takes his life and one gets his life back, BP decides that an American, Bob Dudley, will replace Tony Hayward. Dudley spent much of his childhood in Mississippi.

Various people in the news media continue to complain about hassles with BP's private guards and about cops and sheriffs' departments doing BP's bidding in clear violation of public rights of freedom of movement on public property. Weeks ago, on June 6, Thad Allen had told ABC News, "I put out a written directive and I can provide it for the record that says the media will have uninhibited access anywhere we're doing operations, except for two things: if it's a security or safety problem. That is my policy. I'm the national incident commander." So there you have it. He's the decider.

The memo, signed by Allen on May 31 and sent to various government entities and to BP, says: "In any matter whatsoever, and at any level of the response, the media shall, at all times, be afforded access to response operations and shall only be asked to leave an area when their presence is in violation of an existing law or regulation, clearly violates the written site safety plan for the area or interferes with effective operations."

Allen's memo says, "the media." I'd have preferred it to also say the public, because people who work for conservation groups, scientists, book writers, freelance or part-time photographers, fishermen who know the area, and folks like that don't have media IDs. Later down the memo says, "No contractor, civilian employee or other responder involved in the Deepwater Horizon response has the authority to deny media access to operations except as noted in paragraph one." That's pretty clear. And the intent is clear.

Okay, great. Now let's see the directive in action. On June 22, weeks after Allen's "uninhibited access" order, *Mother Jones* magazine's website posts a video of a "law officer" hassling a guy from the American Birding Association for filming the exterior of a BP office building from across the street. Andrew Wheelan was not on BP's property at the time, but the law officer nevertheless carries out BP's intimidation program:

Wheelan: "Am I violating any laws or anything like that?"

Guard: "Um . . . not particularly. BP doesn't want people filming."

Wheelan: "Well, I'm not on their property so BP doesn't have anything to say about what I do right now."

Guard: "Let me explain: BP doesn't want any filming. So all I can really do is strongly suggest that you not film anything right now. If that makes any sense."

Let's make the rapid trip from no sense to incensed: Shortly thereafter, Wheelan got into his car and drove away, but he was soon pulled over. It was the same cop, but this time he was with a guy whose badge read "BP Security." The cop stood by as "BP Security" interrogated Wheelan for twenty minutes, asking him who he worked with, who he answered to, what he was doing, why he was down here in Louisiana. Mr. BP Security phoned someone and, just to be mean, confiscated Wheelan's bird-helper volunteer badge. Eventually, he "let Wheelan go." But bear in mind, this is a private security guard, pulling a citizen off a public road.

It gets better. "Then two unmarked cars followed me," Wheelan says. "Every time I pulled over, they pulled over." This went on for twenty miles.

Bye-bye, God bless America; hello, corporate police state. So easy. And no blood.

Coda: the "law officer" was an off-duty sheriff's deputy for Terrebonne Parish. Off duty—and working in the private employ of BP. The deputy failed to include the traffic stop in his incident report. In other words, he abused his authority and then hid the fact. But he had support from higher up: a major in the sheriff's office tells the magazine's writer, Mac McClelland, that an off-duty deputy using his official vehicle to pull someone over while working for a private company is "standard and acceptable practice" because—get ready—Wheelan *could have been a terrorist.*

Of *course* he could. It's not like BP is at the epicenter of a giant oil blowout and someone with a video camera might want to post an image of BP's headquarters, with its huge logo, on the Web, or anything like that. It's much more likely that a terrorist would be standing across the street with a video camera and a magazine writer in broad daylight, talking to BP's guards.

Ergo, back to Thad Allen: "The media will have uninhibited access anywhere . . . except . . . if it's a security or safety problem." And it's *always* a security or safety problem—because that's all they ever have to say to do anything they want.

It could be a cleanup; it could be a cover-up. You can't tell. You can't tell because the Big People are undermining our ability to ask. But let's make it simple, people: Either there's freedom of speech or there isn't. Either there's freedom of assembly or there isn't. Either there's freedom of movement or there isn't. Either there's freedom. Or not.

And what there is here and now is: bullying and lying at the speed of sound. Illegal, sure; but when law enforcers agree, it gets very hard to deal with. Why they agree, why they get turned against the public, I'm not sure. Something about the liberal media? Something against outsiders? Boredom and a chance to throw their weight around with impunity? It's all a little dose of "the banality of evil" (a phrase originally coined in reference to the Nazis). The idea: it doesn't take terrible people to do bad things in an official capacity. It takes average people. Average people who want to do a good job for their superiors, want to be loyal, who know how to go along to get along, and who like to avoid any risk to themselves. Unfortunately, that's all it takes. Average people.

Who are the "terrorists"? Who are the ones acting against America's principles? The people who don't want you to see the pictures, or those who do? The people who abuse their authority, or the ones they abuse? The people whose reckless rush risked hurting all "the little people," or the people with little, who stand tall?

And here's the main thing: even if the Coast Guard has taken the spirit of Allen's media-access memo to heart and fully embraced his directive (I said "if"), BP and local law enforcers are ignoring it. They're doing whatever they want when they feel like pushing people around. And this is America. These companies are multinational. Imagine what they do elsewhere.

What they do: In Nigeria, an amount of oil roughly equivalent to that lost by the *Exxon Valdez* spills into the Niger Delta every year. It has destroyed farms and forests, contaminated drinking water, driven people from their homes, and ruined the nets and traps of fishing people. On May 1, 2010, a ruptured ExxonMobil pipeline spilled more than a million gallons into the delta over seven days. Local people protesting say security guards attacked them. Thick tar washed ashore along the coast. Said Bonny Otavie, a member of Parliament, "Oil companies do not value our life; they want us to all die."

Nigerians can scarcely believe the efforts to stop the Gulf oil leak and to protect the Gulf shoreline. When major oil spills happen in the Niger Delta, Nigerian writer Ben Ikari observes, "The oil companies just ignore it." The Nigerian government says there were more than 7,000 spills between 1970 and 2000. Nearly one-tenth of the oil America imports comes from Nigeria.

Shell says that 98 percent of all its oil spills in Nigeria are caused by vandalism, theft, or sabotage. Local communities and environmental groups insist that the problem is rusting facilities and apathy. Similar stories come from the Amazon, Ecuador, and elsewhere. Nigerian environmentalist Nnimo Bassey says, "In Nigeria, they have been living above the law. They are now clearly a danger to the planet."

By the start of the third week of June, one-third of the Gulf's federal waters, 81,000 square miles, remain closed to fishing.

And the *Economist* estimates that BP is on the hook for eventual cleanup costs and damages of $20 billion, plus fines up to $17 billion. But BP's market-value plummet is two to three times as great. BP stock has melted off nearly $90 billion worth of value. Investors fear that compensation claims are spiraling out of control. I think they're overreacting. BP, I am willing to say, will probably be fine.

Big Oil has long enjoyed the milk and honey of tax-fed privileges ranging from massive subsidies to supreme dispensation. Exxon led the small communities of Prince William Sound in a grimly choreographed death dance that ended in 2008. When a huge penalty was levied against Exxon, the oil giant got the U.S. Supreme Court to hear its case nearly two decades after *Valdez* ran aground. Chief Justice John Roberts began his inquiry by asking, "Isn't the question here how a company can protect itself from unlimited damages?"

No, John, that isn't the question. The question is: how can *people* be protected from unlimited damage?

A jury had awarded $5 billion in damages, but the Bush-wacked Supreme Court said, "No, it'll be more like ten percent of that." Thanks to the antisocial, pro-corporate ideology of certain "justices" still seated on the Court for life, the oil titan paid just $507 million (10 percent) of the $5 billion damage settlement that a lower, better court had arranged, and just $25 million (17 percent) of its $150 million initial fine. The payments were a tiny, momentary blip on Exxon's profit spreadsheets—and a second catastrophe for the real lives of real people of actual communities. Nineteen years after the *Exxon Valdez* ran aground on Bligh Reef, some plaintiffs received their final payment. Others had already died.

And today the Court remains stacked with Bush appointees as thoughtless, more heartless, and more pro-business than it was then.

The aftermath of the *Exxon Valdez* spill—its devastating effect on the region's wildlife, its long-lasting depression of fish prices, the social and economic strains that followed, Exxon's antisocial behavior and the way the Supreme Court swam with it into the toilet—set the bar so low, it's as if someone dug a trench and threw the bar in. One nation under oil. The *Exxon Valdez* spill was more than a tragedy, more than a crime. It remains a national stain and a national trauma. The fear that this Gulf blowout will be "as bad as the *Exxon Valdez*" will remain in hearts, on minds, and on lips throughout.

Back on May 6, President Obama had declared a three-week moratorium on exploratory oil drilling in the Gulf of Mexico, to give his administration time to review safety regulations and the quality of government oversight. While drilling technology has exploded—poor word choice—*improved* incredibly in the last thirty years, allowing location and extraction in ever-deeper, harder-to-reach regions, cleanup equipment has gone nowhere. Since the Ixtoc blowout, since the Exxon spill, it's the same old booms, skimmers, dispersants, and guys with shovels. They make money from oil, so they put money into oil. They don't make money from cleanup, so they ignore cleanup. Big mistake, because accidents can be costly—but it's obvious how little they've cared.

Nearly two hundred miles from shore, a $3 billion floating oil platform much larger and more complex than the Deepwater Horizon straddles the deep ocean like a giant steel octopus. Named Perdido (Spanish for "lost"), this colossus pumps oil from dozens of wells in water nearly two miles deep—while simultaneously drilling new ones. The pipelines flowing from wells to rigs like this can be tens of miles long. Compared to this monster, the Deepwater Horizon was a simple little rig with the luxury of focusing on one task in relatively shallow water. Meet the new wave: *ultra-deep* platforms.

It's safe, of course. Accidents are rare, as we've seen. And also, of course, the stakes—and the risks—increase as the rigs get larger and more complicated.

"Our ability to manage risks hasn't caught up with our ability to explore and produce in deep water," says Edward C. Chow, a former oil executive now with the Center for Strategic and International Studies. Perdido, for example, lies twenty hours away from supply-boat help. It lives in a realm of hundred-mile-an-hour hurricanes and mountainous walls of angry, battering-ram waves. Its delicate underwater equipment and many pipelines feel the insistence of currents and mud slides, but lie far beyond human reach in their own underwater metropolis populated by unmanned submarines and robots. Down there, it's *Dune*—but without people.

In 2005, Hurricanes Rita and Katrina damaged or destroyed hundreds of offshore platforms and pipelines. (About thirty thousand miles of pipeline crisscross the Gulf of Mexico seafloor.) The gulf's oil and gas production shut down for weeks. I hope we don't someday look back at that as a quaint time of heroic people. Meanwhile, new rigs have recently arrived in the Gulf that can drill in water 12,000 feet deep.

"Going to the moon is hazardous. Going to Mars is even more

hazardous," says University of California professor Robert Bea. "The industry has entered a new domain of vastly increased complexity and increased risks."

On May 27, Obama had extended the ban on deep exploratory drilling for six months. Today, June 22, a federal judge strikes down the moratorium, saying, "Are all airplanes a danger because one was? All oil tankers like *Exxon Valdez?* All trains? All mines? That sort of thinking seems heavy-handed, and rather overbearing."

Of course, that's not the point. When a plane crashes, hundreds of thousands of people don't get put out of work, nor do they perceive their communities, livelihoods, and self-identities threatened; an entire region doesn't lose tens of billions of dollars. That's a difference. The judge says the six-month moratorium would have an "immeasurable effect" on the industry, the local economy, and the U.S. energy supply. Maybe *he* can't measure it, but economists should be able to tell us how many jobs, what overall effects, things like that. All those big robes and that big bench don't guarantee much. Beware the man behind the curtain.

And yet, I have to admit that even I'm not sure that the Obama moratorium is necessary. It seems that regulators could greatly improve rules and tighten oversight without a moratorium. But I just don't like the judge's juvenile logic. His silly comparisons are off base. Rather than blind justice at work, I see a certain blindness. Like most everyone in this mess, he doesn't grasp the big picture: this isn't like a plane crash; it's like aviation safety procedures. There are systemic problems to fix.

In response, the Interior secretary says he will order a *new* moratorium, one designed to eliminate doubt that it is appropriate. Salazar says he expects oil companies to complain that the coming regulations are too onerous. "There is the pre–April 20th framework of regulation and the post–April 20th framework," he'll say, "and the oil and gas industry better get used to it."

Despite the judge's ruling, BP stock drops 2.7 percent. The better bet: shrimp. Imported shrimp prices are up 13 percent, and Thai shrimp import poundage is up 37 percent. And Gulf shrimp prices have gone jumbo, up a whopping 43 percent. About 60 to 70 percent of oysters eaten in the United States come from the Gulf, and oysters now cost about 40 percent more, too.

On June 23 the feds reopen fishing over more than 8,000 square miles. Their reason: "because no oil has been observed there." So why'd they close it? Other parts of the Gulf with no observed oil were never closed. Still closed: roughly 75,000 square miles, about 32.5 percent of the Gulf's federal waters. Their smiley face: "That leaves more than two-thirds of the Gulf's federal waters available for fishing." People have now found more than 500 sea turtles dead or dying around the Gulf.

During the last days of June, Tropical Storm Alex thickens the air with sheets of rain that pound so heavily on my car that several times, unable to see, I have to pull over. Seven- to ten-foot seas lash the coast. Beach cleanup is halted, most work disrupted, most people scurrying for shore; hundreds of vessels steam for ports. Waves are pushing most booms ashore. Movement of the surface oil accelerates; winds push the crude turds northwest, toward heretofore oil-unsoiled Texas beaches.

Thad Allen says the weather could suspend operations for two weeks. It "would be the first time and there is no playbook," he says with dramatic flourish. No playbook because the BP Gulf plan mentions walruses, but doesn't mention tropical storms or hurricanes.

One of the bigger worries: if they have to make the capture vessels disconnect from their supply lines, all that oil just resumes leaking full-on into the sea. Upgraded to a hurricane, Alex is the strongest June storm since 1966, with sustained winds of more than one hundred miles an hour. But its main winds pass wide of ground zero, and the rigs stay connected. The winds, though, splatter oil onto the beaches of South Padre Island, Texas, and Galveston.

On June 30, 2010, every Republican in the House of Representatives votes no on a bill that would require corporations to disclose the money they give to American elections.

Also on June 30, Coast Guard admiral Thad Allen becomes retired Coast Guard admiral Thad Allen. He joins the Department of Homeland Security and will continue managing the federal oil spill response.

Meanwhile, the widows of workers killed in the Deepwater Horizon explosions are being told that Transocean plans to argue that its liability for damages owed is limited by the Death on the High Seas Act and the Jones Act. Shelley Anderson, whose husband, Jason, was a tool pusher on the rig, says, "Why would the damages to a family be different if a death occurs on the ocean as opposed to on land?" Well, Ms. Anderson, it's not that the damage to your family is different. It's just that, having caused your husband's death, the corporation doesn't want to pay you. That's just the kind of people their executives are. Remember, no belly button; they've got their wallets in their chest pockets, where their hearts should be, and are bereft of pulse.

In the true heart of the delta, what land there is lies like giant snakes resting in shallow water, each snake just wide enough for a road, a few docks, some homes. That's all. The people and communities seem as aquatic as muskrats.

Things have changed in the last two months. When I first came, people were in shock. Now most people have worn a slight groove in their situation.

The first place I'd visited was Shell Beach, Louisiana. So now I'm going back. This time I have a little company for a change. Mandy Moore works for the National Wildlife Federation. Blond, twenty-nine, slight of build, she's from around here, knows the place. So she's driving. I'm in the passenger seat eating peanuts and some cherries.

"We won't go to the staging area," she says. "They won't let us down that road at all. That part's gotten a lot worse."

Other road. Another sheriff booth. We'll park before it, plan to walk through. Our calculus on their psychology is that walking is less confrontational, somehow, than driving on a public road. That's how distorted things are. Mandy tells the guard we've gotta talk to Frank Campo. Dropping a name earns us permission to walk on a public road.

Just a little bit down the road is an open-air fishing station with a corrugated aluminum roof, a dock, tanks for live bait, fuel pumps. Fishing's been closed for a couple of months.

Campo's ancestral ethnicity is Spanish via the Canary Islands to Louisiana. His fishing station normally sells bait, fuel, and ice to commercial and recreational fishermen. He's spent a lot of time in the sun.

Hurricane Katrina sent twenty feet of water through here. "There's only two things here that's old, besides me," Campo says, pointing, "That post, and a piece of pipe in the front. The rest, Katrina destroyed. The whole parish was destroyed. You couldn't buy anything. You couldn't get here to work because the road was destroyed. All them power poles, they were all gone. They never rebuilt the gas lines, not for the few people here. We knew we were gonna rebuild, but the question was 'Where the hell do you start.' My dad was dead. But I asked him, 'Dad, where the hell do we start?' And we talked about it and he said, 'We have to rebuild; this is what we do.'" It took about a year to get back in business. Campo's son Michael, late thirties, is fourth-generation in this family business. Michael says he's "tryin' t' stay positive." So much of this is psychological warfare.

Campo the elder says, "When I first heard about the oil, to be quite honest, I didn't think much of it. Rigs blow up all the time, y'know. Then I saw that eleven people lost their lives. That really bothers me. That's not good. You got kids? I sure wouldn't want to be rubbed out because somebody did something stupid. I mean, I got grandkids and I know I'm not gonna be here forever; but I wanna be here, y'know, as long as I can. So I feel really bad for the people who lost their husbands and their fathers.

"Katrina destroyed us—but it didn't kill us. A hurricane takes everything, but you know you're gonna come back. You know you're gonna have the seafood, sport fishing—. I mean, it takes a while, but you know you're gonna be back on top of the ball again. But the oil really bothers me. The oil could take all this away from us. What do you do then? And *where* we gonna go? What's the use of coming here if you can't fish?

"I *fish*, y'know what I mean? Louisiana produces most of the seafood that's eaten in the country. They got shrimp comes outta other states but they ain't no good; I wouldn't eat 'em. Our shrimp are *so* good, it's not even funny.

"I've fished out of so many places, I got friends all over the doggone country. Biloxi, Gulfport, Houma. I've been all over. I got friends in Texas. I didn't have to punch a clock. And I didn't have to drive. I went with the boat, everywhere.

"This fishin' is a learnin' process. You get to meet interesting people. And you learn a lot from everybody. When I started, you used a trawl. Well, all right. Then they started using what they call a tickler chain ahead of the net. It makes the shrimp jump up. Well, that improved the catch—greatly.

"As we travel through the jungle, you gotta change with the times. But now—. I don't know *what* we're gonna do. Because if the oil moves in, it could kill the marsh. I don't feel confident we'll survive this. This is a significant threat to our well-being. This is not something to take lightly. This is very serious. This could destroy a way of life for everybody in this part of the world. If they don't stop it, then we'll be dead. Eventually, this is gonna go away. Whether it's going to take everybody with it, that I don't know."

Over on the *Ellie Margaret* we find Charlie Robin—"S'posed to say Ro-*ban*," he acknowledges, "but it's Robin." His boat, designed for shrimping, is laden with that elongated pacifier of the Gulf summer of 2010: boom. But Robin isn't quite pacified yet. He's mad as a buzzworm. Angrily, but amiably—he knows who his enemies are *not*—he sits astride the side of his battered boat. Not a lean man. Talkative. Worried. On top of it all, mechanical problems. Hit some leftover hurricane debris. Bent the propeller. "Destroyed it." Broken exhaust. Clothes covered in grease. Hands full of grease. Last week, he cut his finger off in a winch. "Dat little-bitty winch right there? Cut it off." Reattached, heavily bandaged.

Gesturing to the boom on his deck, he says, "Ninety percent of the boats here are workin' de erl. Nevvah thought it would still be comin' out in July. . . . Gettin' my flares. Gettin' all my safety equipment. Dat's what I'm doin'. Dat way we's good t' go. We been practicin' in a lake. We pull two twenty-four-inch booms. And we tow at, like, one knot. You got t' go a certain speed. Don't want the erl to go over, don't want it under. So you go *slow.* Other boat's behind you; he skims it. Pumps it on a barge. That's the way you go. Mop it up. Exactly what we doin'. So, pretty cool. I'm excited about it. Because I know it's gonna help our fisheries out. Anything we do to save our land, save our area, it's good. I'm five generations. If I go shrimpin' and de erl comes in, I'm screwed, right?

"We don't *know* what's the effect all this gonna be. Dat's de scary part. Dat's the part we feah de mos'. *No one* not knowin'. Hurricane comes in, we clean up, lick our wounds, go back workin'. We might bust a few nets on debris, but we back to livin'. But this?" His voice drops to an emphatic whisper: "We don't know the *outcome* of this. We don't know, we don't know.

"The ground you standin' on been in de family a hundred and fifty yeauhs. My great-grandfather lived on it. It means a *lot* to me. And it's worth savin'. This place goes, what we gonna do? Is BP gonna give us back owah culture? *Hell* no. There's not enough money in the world to pay for five generations of freedom. You don't *buy* dat.

"And to give you fresh seafood, Mother Nature's home brew—instead of some farmed frozen foreign chemical-raised shrimp. You follow dat? Okay.

"Well, *dat's* how we feel. Dat's how *angry.* Their ignorant mistake—is gonna cost all this. To try to save a million bucks. Dat's a penny to us. Now I'm gettin' mad. I'm sorry."

As we're walking out, back up the road to Mandy's car, we pass a handful of young people hanging out, leaning against a parked car. One, in his early twenties, is wearing the little orange life vest of the body snatchers. (He's on land, mind you—safety!)

Just after we pass he yells, "Hey. *Hey. HEY!*" Apparently he's "working," but doesn't know the difference between adolescence and a job.

We're walking *out*, on the public road, and he's yelling at us. Mandy has the courtesy to stop, so I turn around, too.

He suspects we're "from the media."

Mandy says we're not "the media."

"Then what's *that*," he demands. She happens to be carrying a folder bearing the business card of a *Los Angeles Times* reporter. Seeing that, the guy believes he's caught her lying to him. These are the banal ways the Gulf becomes a police state.

As if he deserves any answer, as though he has any scrim of real authority, Mandy calmly tells him that, no, it's someone else's business card. It's a good thing he's talking to her (that's an easy choice for him, of course) and that she's doing the talking, because I can feel my next breath forming the words *Go fuck yourself.*

⁂

As oil speckles Mississippi beaches, Governor Haley Barbour complains of not being given adequate resources. A Democrat congressman says he's "dumbfounded by the amount of wasted effort, wasted money and stupidity that I saw."

Oil spreads. Pain deepens. "Seeing everything that you've been used to for years kind of slowly going away from you, it's overwhelming," says a boat captain. His family wants answers he doesn't have. His wife says she cries. A lot. "I haven't slept. I've lost weight," says Yvonne Pfeiffer, fifty-three. "The stress has my shoulders up to my ears."

Oil floats. Shares sink. BP's stock price drops 6 percent, to $27.02, on June 25. Lowest value in fourteen years.

Oil floats. Turtles fly. The U.S. Fish and Wildlife Service is coordinating collection of about 70,000 sea turtle eggs from around 800 beach-buried nests from the Florida Panhandle to Alabama. The fear:

hatchlings will head into a toxic sea and be fatally mired as they seek food and shelter in oil-matted seaweed. The dilemma: nobody's ever done this before at such a scale. There's a lot of guessing. And any hatchlings that survive will almost certainly not return to the Gulf as adults, because they'll imprint on the east coast of Florida beaches where they'll be released. The turtles could squeeze out a win. But for the Gulf, it's a loss either way. It's not that anyone thinks this is a great idea. It's that the people involved don't want to just sit back; they want to help.

Halfway around the world is a man who wants so very much to help, he has sent us a giant, unsolicited gift. And so now, in the center ring, we have for you: the World's Largest Skimming Vessel. That's right, ladies and gentlemen. This massive beast, the awkwardly named *A Whale*, has crossed the Pacific flying a Taiwanese flag to go head-to-head, mano a mano, with the slick. How many football fields is it? you might ask. Three and a half! Ten stories high! It's a tanker newly converted at its owner's expense for just this purpose. And what a guy he must truly be. It has never been tested, but it's the thought that counts. And this whole season is about testing technology and people as never before.

The radio theater of the absurd tells us that "officials" hope the vessel will be able to suck up as much as 21 million gallons of oil-fouled water per day! But that's not quite true. That's what the *owner* hopes. Remember, it hasn't been tested. And the key phrase there is "oil-fouled water." Nobody thinks it could collect that much sheer oil.

Officials are doing their usual bit: being skeptical and mulling whether to grant access. And so the ship's mysterious and "reclusive" owner has sent a representative from a PR firm to "unleash a torrent of publicity and cut through red tape."

Turns out, "officials" are inclined to deny the ship permission to work, citing the fact that after it skims water and separates out the oil, the water it returns will still have some oil in it—*and it's illegal for a ship to discharge oil.* Never mind that there are skimmers all over the

Gulf right now doing just that. "BP and the Coast Guard still have concerns about the ship," one official says.

I continue to have a concern: Why does it keep sounding like "BP and the Coast Guard" are one team? How is it that BP gets to say *anything* about what and who goes into the federal waters of the United States? Especially when we the people of the United States sit helplessly watching BP fill the Gulf to the brim with oil and dispersants?

Finally, finally, finally, after about a week, the giant skimmer ship is allowed to do its thing.

It doesn't work.

On the final day of June, oil is scoring a touchdown on the Mississippi coast. On the radio, Governor Barbour says they had a great plan six weeks ago that isn't working very well now. Not enough skimmers, not enough equipment.

He is asked by the radio host whether this tests the philosophy that he and many Republicans champion: smaller government, less regulation, more freedom for industry. "The idea that more regulation is good is, I believe, a very suspect idea," Barbour answers. "In the case of this well, I think that if existing regulations were followed, it wouldn't have blown out. I think if there was somebody from Minerals Management Service on the rig that day making sure the regulations were properly followed, that would've made a difference. Now I think every oil company in the world is looking and thinking they wouldn't want to be paying $100 million a day like BP. That's how I believe the market system works."

And why *wasn't* the "Minerals Management Service on the rig that day making sure the regulations were properly followed"? Precisely because of people who champion smaller government, less regulation, more freedom for industry. Let's face it: most people who think they're "conservatives" these days are mainly phonies, radical front people for big business dedicated to removing public safeguards and safeguarding private greed. The missed point about whether government should be small or big, strong or weak is: it's *our* government. It should be accountable *to us.* Real conservatives would tell corporations to go to hell when they try to contribute campaign money, when they work to influence elections. By allowing themselves to become obsessed with the demand for "small government," "deregulation," and taxes, in effect they mostly represent big corporations that pocket profits and dump their risks and the costs onto other people. Some of the other people are too angry to realize that there's a shell game going on. The rest of us are simply too comfortable. "The best lack all conviction," Yeats said, "while the worst are full of passionate intensity."

It's hot. And because it's so hot, BPs beachside cleanup workers—30,000 of them—are told to work for twenty minutes and rest for forty. For $12 an hour, the work is sweaty and uncomfortable, but not overly taxing. To save their backs, workers are not allowed to put more than ten pounds of oily sand in a bag. That's not much sand. Hundreds and hundreds of plastic bags, each with its little dollop of the besmirched beach—where are the hundreds of thousands of plastic bags, oily absorbent materials, and hundreds of tons of oily trash going? Landfills. Mixed with regular trash. Because it has to go somewhere. Don't worry: BP says that "tests" have shown that the material is not hazardous.

Someone on the radio is saying, ". . . To get near oil on the beaches, people need special training. But BP has troops . . . *thousands* of workers, who have received such training. . . ."

They make oil sound so *special.* It's *oil.* Gasoline's more dangerous. Ever fill your car's tank? Ever change your car's oil? You don't need special training. For anyone with a pair of old sneakers and a shovel, it should be no Big Problem to pick up a little oil if they want to.

Out on the water, some of the workforce sits idle. About 2,000 vessels are supposedly involved in the cleanup efforts. But many captains sit aboard their boats, awaiting instructions. For this, BP pays them, say, $1,000 a day (the fee varies with the size of the boat). Meanwhile, thousands of people from around the country want to drop everything and come help. But there's nothing for them to do. And their calls don't get returned. Forty-four nations have offered to help. For the most part, their calls don't get returned, either.

"The clean-up effort has not been perfect," BP's new American spokesmouth, Bob Dudley, acknowledges. But, seeking to assuage fears in the only language known to multinational corporate brains, he adds that BP remains a "very strong company in terms of its cash flow."

Science bulletin: an astonishing congregation of dozens of the world's largest fish—whale sharks—have been discovered in the Gulf of Mexico. One aerial photograph showed about ninety of the behemoths together, about sixty miles from the oil. "It blew my mind," says the University of Southern Mississippi's Eric Hoffmayer. The bad news? You guessed it: some of them have also been seen in heavy oil. They eat tiny creatures that they strain from the water. Their feeding technique includes skimming the surface and moving almost 160,000 gallons of seawater through their mouths and gills per hour as they feed on tiny fish and plankton. Watching them, you get the impression that their feeding method is the worst possible technique for surviving an oil slick. "This spill's impact came at the worst possible time and in the worst possible location for whale sharks," Hoffmayer says. "Taking mouthfuls of thick oil is not conducive to them surviving."

Science bulletin: scientists with the University of Southern Mississippi and Tulane University report finding petroleum droplets on the fins of small larval fish. "Their fins were encased in oil," says Harriet Perry, director of the Center for Fisheries Research and Development at the Gulf Coast Research Laboratory. "This is one route up the food ladder," she speculates. "Small fish will eat the larvae, bigger fish—you know how it goes."

Laboratory studies and field experience with wildlife shows that oil can cause skin sores, liver damage, eye and olfactory irritation, reduced growth, reduced hatching success, fin disintegration, and, of course, death. In heavily oiled areas, some animals can move away. Fish eggs and larvae cannot move. The *Argo Merchant* oil spill killed 20 percent of nearby cod eggs and almost half of pollock eggs. The *Torrey Canyon* killed 90 percent of the eggs of a fish species called pilchard. Similar death rates followed the *Exxon Valdez.* Other species seem far less affected. Of oiled fish eggs that hatched, larvae often had deformed jaws, spinal problems, heart problems, nerve problems, and behavioral problems.

However, under normal circumstances in clean water, natural mortality rates of eggs and larvae are so colossal, and such a tiny fraction survive to adulthood, that even the near-total destruction of one whole year-class of fish eggs by an oil spill might have a difficult-to-notice effect on adult populations.

In Prince William Sound during the months following the *Exxon Valdez* spill, herring eggs and larvae in oiled areas died at twice the rate they did in unoiled areas. Larval growth rates were half those measured in other North Pacific populations. Herring larvae also suffered malformations, genetic damage, and grew slower than ever recorded anywhere else. Those problems were gone by the following year.

But it's not that simple. Different things get hurt at different rates for differing periods of time. Oil that works its way into sediments and under boulders remains toxic and available to living things. In Prince William Sound after *Exxon Valdez,* oil hiding beneath mussel beds continued to find its way into the region's animals and their food web. For years, ducks and otters suffered chronic exposure to oil. For at least four years, the eggs of pink salmon, which spawn in the lower reaches of streams, near seawater, failed at abnormally high rates. Young sea otters born for several years after the spill survived at unusually low rates. After sea otters had received protection from hunting for their fur, their population increased 10 percent annually. But after the oil spill, they recovered at only 4 percent per year. In heavily oiled areas, their numbers did not increase at all for at least a dozen years. Shellfish—which sea otters (and people) eat—concentrate oil hydrocarbons quickly and metabolize them slowly. For at least several years, black oystercatchers fed their chicks more mussels but achieved less growth than normal. Harlequin ducks (probably the world's most exquisitely beautiful sea duck, which is saying something) for many years suffered low weight as their bodies tried to fight the toxic effects of the petroleum hydrocarbons they were getting in their food. In parts of Prince William Sound, they died at rates of 20 percent annually for over a decade. A study published in the April 2010 issue of *Environmental Toxicology and Chemistry* finds that harlequin ducks are still ingesting *Exxon Valdez* oil. Biopsy samples show their livers containing the enzymes they produce when their body is wrestling with oil.

Everywhere I've been, there's boom. Boom along the shore, boom under bridges, boom in roadside canals. Boom, boom, boom. You can't avoid it. And almost everywhere I've been, there's been the chronic low-level hassling of camera-toting types like me, by sheriffs and orange-vested private guards on public roads and alongside waterways.

And today, the Coast Guard takes the situation one giant step in the wrong direction. They make it a crime in southern Louisiana to get within seventy feet of boom. You risk a $40,000 fine. And if you do it "willfully," that's now a felony.

A *felony?* Impossible. Can the Coast Guard *make* a law? But it's true. Here is their press release:

> June 30, 2010 16:51:40 CST
> Coast Guard establishes 20-meter safety zone around all Deepwater Horizon protective boom operations taking place in Southeast Louisiana.
>
> The Captains of the Port for Morgan City, La., New Orleans, La., and Mobile, Ala., under the authority of the Ports and Waterways Safety Act, has [*sic*] established a 20-meter safety zone surrounding all Deepwater Horizon booming operations and oil response efforts.

> Vessels must not come within 20 meters of booming operations, boom, or oil spill response operations under penalty of law. . . .
>
> Violation of a safety zone can result in up to a $40,000 civil penalty. Willful violations may result in a class D felony.

This, after weeks of people screaming for transparency and complaining about interference and petty bullying by people getting paid by BP. When I do a Google search with the words "media access Gulf oil," I find plenty of other people complaining. One Web commentator says, "Never in my lifetime could I imagine that a foreign company could dictate my ability to move freely and openly in American territorial waters."

America should be able to show that in a crisis we are at our finest, our most American. But wow. I can barely contain the rage I feel at the Coast Guard and its Thadmiral.

The highway to Venice, Louisiana, is sixty miles of levee sandwich, a corridor of road between corridors of water. Heavy shipping lanes gouged through wetlands. Dying trees in subsiding marshes. Herons. Cormorants. A least bittern; nice bird. They're all nice. Egrets still immaculately white, offering the hope of the living even as they feel the squeeze.

Captain Jeff Wolkart is telling me, "Two weeks into the spill, we were at Pelican Island fishing under birds in about five feet of water, which is a common way of fishing this time of year. And a dolphin kept coming around. Its body was covered in that brownish oil, that tannish-colored crude. And he was trying to blow out his blowhole, and he was struggling. Porpoises scare fish, so I moved off a hundred yards. It followed. I kept doing that and it kept coming back, coming to us, hanging right alongside the boat. That's very unusual. They're pretty intelligent, and it seemed to want help. But eventually I had to leave."

Dawn. Helicopters soon join the gulls. The drone of engines is as incessant as the industries of swallows that affix their nests to the I-beams of waterfront warehouses.

The media invasion is pulling away, leaving a low-level occupation. So many were they—many of them urban northerners—that the dockside burger-and-fries joint I'm in was compelled to add a "healthy platter." BP, its contractors, its security guards, and more than a few sheriffs understand the media as spies, prying eyes for a public prone to unwieldy concern if well informed.

The watermen seem a bit severed by the constant traffic of contractor vans and trucks and the private guards. It's not their place anymore. How are they finding the strength?

Here's how: all the watermen are choosing to work for BP. There's money now, work moving stuff, bringing stuff, towing stuff. Two grand a day. For now. After that, they don't know.

Charter captains meeting: fishermen—former fishermen, for now. A young man, twenties, asks, "Where will I take my kids fishing?" He has no kids yet. I hope he becomes CEO of an oil company. They could use his concern for kids.

All these guys know what it was like a couple of months ago, before the Big Problem. Few know, as the head of this association knows, what it was like way back. "You young guys, you'll never see it like me and Billy saw it," he says. I follow his eyes to where Billy's gray hair flows from under a baseball cap. "The place is a small fraction of what it was. It's infinitesimal compared to what it was."

The place is about water, cane, tides, mud, crabs, fish, birds, shrimp. All the pesky things an engineer overlooks while making "America's Wetland" bear the tasks of shipping, industrial access, flood control.

This isn't just a working coast. It's an overworked coast. Watermen feel stalked by disintegrating marshes. And everywhere the towering hardware of oil and gas prickles the horizon and brings down the sky. Henry Ford first used biodiesel. Standard Oil lobbied for Prohibition so there'd be no ethanol available. Ford was forced to switch to petroleum gasoline.

Outside, the oil complex confronts the eye with pipes, stacks, tubes. "It's so ugly," my companion says, "but it's jobs." A moment later she adds, "Ports, rigs; the oil is just the face of how the whole place—nature and people—have been so disrespected."

LIKE A THOUSAND JULYS

By the beginning of July, this blowout achieves peerdom, in sheer volume, with the Ixtoc disaster. That had been the largest accidental release of oil ever. Until now. Something like 140 million Macondo gallons have hemorrhaged into the Gulf.

Way back on May 26, the Environmental Protection Agency ordered BP to cut dispersant use roughly 75 percent *from the maximum.* But the maximum was 70,000 gallons in one day. The EPA now says it wants to keep dispersant use down to 18,000 gallons per day. CNN reports that BP is still averaging about 23,250 gallons. Such are the games.

EPA administrator Lisa Jackson said weeks ago that she was "dissatisfied with BP's response." Her agency set a deadline for BP to stop using two particular Corexit dispersant formulations (including one banned in Britain).

In mid-June BP announced that one, Corexit 9527, is "no longer in use in the Gulf." The manufacturer will say that the alternative, Corexit 9500, does not include the 2-butoxyethanol linked to the long-term health problems of *Exxon Valdez* cleanup workers.

BP seems to get away with shrugging its shoulders. When it only partially complies, there are no fines and no one goes to jail. I know what would happen to, say, me if I took a boat into the Gulf and radioed the Coast Guard to announce that I was about to dump one barrel of chemical dispersant.

And at the beginning of July, more than two months into the blowout, the Environmental Protection Agency comes out with new findings on the dispersants: They're "practically non-toxic."

Because of the early "not leaking; too early to say catastrophe" statements by the Coast Guard, because of the tone implied by "anomalies, not plumes," because of the BP–Coast Guard marriage and the chronic official bullying, because the EPA seems to let BP shrug off even weak directives, because, because, because—nobody believes "them" when they say the dispersants are "practically" non-toxic. Whether they're right or wrong is beside the point. The point is: "they" have a credibility problem; they've lost the people's trust. No one is sure who to believe. There is no real source of reliable information, and people continue to believe, think, and feel a lot of things that aren't true. But some are true. It creates confusion.

So, rightly or wrongly, people remain concerned that the dispersants will kill plankton, fish, and other marine life. People are concerned they'll taint seafood. People are concerned that they'll break up millions upon millions of gallons of oil so that it pollutes vast swaths of the Gulf, making the oil unrecoverable, rendering millions of feet of boom useless on principle, bathing fish eggs and larvae, making things die, getting long-term into commercial fish, shrimp, and oysters, and taking bolt cutters to the food chain.

Certainly all that is "true" in a strict sense, but the important questions are: How much, for how long? Will it kill a lot of plankton? Will it seriously taint seafood? Or, in the vastness and chemical complexity of the Gulf, will the effects be trivial and temporary? It's so hard to know.

An EPA spokesperson: "Before making any alterations to the current policy of allowing heavy dispersant use, we will need to have additional testing of the dispersants plus the oil."

Allow it, then do tests. Isn't that completely backward? They're allowing the whole region to become a laboratory of lives.

St. Tammany Parish president Kevin Davis isn't having any of this. He simply calls on the Coast Guard commander to immediately end use of dispersants. "Breaking the oil down so that it travels underwater and resurfaces is creating a situation where we can neither see it nor fight it," Davis says. "This is not a prudent course of action."

Comedy Central's Jon Stewart tells us that BP has had 760 "willful, egregious safety violations" over the last three years. ExxonMobil, by contrast, has had one. "Exxon," explains Stewart, "could get seventy times the willful, egregious safety violations and still be ninety percent safer than BP."

A few days after the Coast Guard declares "willfulness" a felony, BP generously decriminalizes free speech in America, encouraging workers to talk to the press if they wish to. About that, a shrimp boat captain says, "Yeah, I saw that notice. But our contract still says our people on our boat can't talk to anyone."

With a perceived crisis in oil supply not just from the blowout but also from Obama's drilling bans and moratoriums, I expect a surge in gasoline prices. But the price of gasoline is going *down.* Is this because calls to get beyond oil might galvanize political pressure to move away from fossil fuels and do more for clean energy—so gasoline prices are getting fixed? It's as if, just when we're waking up to fight the bad guys, they give us another shot of muscle relaxant.

Do you remember the "ultimate solution," relief wells? By the first week of July the first relief well is about 1,000 feet vertically from its targeted point of interception with the blowing well—a target, remember, seven inches wide, 18,000 feet below the Gulf's surface. In scale, the drill pipe is like a wet hair extending thousands of feet from the drilling rig, through the ocean, and deep into the seafloor. How can they know where it is, where it's headed, and where it needs to go?

It's very high-tech. To determine the inclination (angle) and azimuth (compass direction), accelerometers and magnetometers send binary pulses through the drill pipe to the drill rig. If the drill bit has strayed, it can be steered back on course by pressure pads that change the bit's direction. Magnetometers sense the seven-inch target of the original well by detecting an electromagnetic field created by an electric current that engineers send through the blown-out well's casing. In addition, devices resembling torpedoes up to 30 feet long bear sensors and processors that measure gamma radiation emitted by rock, the electrical resistance of any fluids within, and even the magnetic resonance of hydrocarbon atoms. Thus drillers know whether they are drilling through sand or rock, or oil and gas.

But no one makes booms capable of effectively corralling oil in open water. Incredible sophistication—and abject stupidity.

And a whiff of an end: the Coast Guard is talking of a new cap in the works. We also hear that the relief well is ahead of its early-August schedule. Don't hold your breath. The Thadmiral says, "I am reluctant to tell you it will be done before the middle of August because I think everything associated with this spill and response recovery suggests that we should underpromise and overdeliver."

Overdeliver? The only thing they've overdelivered is oil and polluted water. That's the part that has vastly exceeded all expectations. (There will not be a relief well in early August. Or mid-August. Or late August. Or early September. Nor by the ides of September.)

In July's first week, BP reports that nearly 95,000 claims have been submitted and the company has made more than 47,000 payments, totaling almost $147 million. So far, total, it has paid out about $3.12 billion.

Meanwhile, the federal government extends fishing closures to cover more than 80,000 square miles, a third of the Gulf's federal waters.

Says a fisherman, "Everything we've ever known is different now. Anything I ever built, it's gone: the business, my client base, my website—." Another says, "In bed, I feel safe. It feels like everything is okay and I'm away from all this. When I get up in the morning, it is just very depressing."

❧

July 2. Headed to Alabama. Downpours overcome by blue sky. Big lofting clouds.

Lots of churches. And their creepy billboards: "Oil Now, Blood Later.—Revelation 8:8." "The second angel sounded and the third part of the sea became blood." What were they smoking? I like my Bible's translation better: "The second angel blew his trumpet, and something like a great mountain, burning with fire, was thrown into the sea." That does sound a bit like the Deepwater Horizon, doesn't it?

Car radio: ". . . If it were up to me, I'd let them try it, because, let's face it, nothing BP and the Coast Guard are doing is doing much good out there."

"BP has received so many suggestions, it's created a *hotline.*"

A *hotline.* Imagine.

A BP hotline operator in Houston asserts that the spill hotline is just "a diversion to stop callers from getting through." She adds, "Other operators do nothing with the calls. They just type, 'blah blah blah,' no information, just 'blah blah blah.' "

BP denies this, saying, in effect, blah blah blah.

Hand-painted signs on utility poles advertise "Snapper," "Grouper," "Flounder," "Shrimp." Bait and tackle every little while. That's all archival now.

Local news: "Mississippi officials have closed the last portion of the state's marine waters. Any fish caught must now be returned to the waters immediately. Officials say tar balls, patties, mousse, and oil residue [such ugly terms] will continue to wash ashore for two months after the oil stops gushing." As Eskimos are said to have many words for differing kinds of snow and ice, the Gulf now has a vocabulary of oil.

The local news also says, "Oil is now building on Mississippi's shoreline, but officials say the major batch of oil is still twenty miles south."

Officials say. Officials say do this; officials say do this; officials say do this; officials say do this; do this. Out. Out. "Officials say . . . Unified Command says . . ."

I say, Unified Command my ass.

". . . all commercial and recreational fishing has been shut down in Mississippi Sound."

Officials say.

National news: "I'm Michele Norris." "And I'm Melissa Block. Computer modeling suggests that Miami and the Florida Keys are actually more at risk from oil than much of the state's Gulf coast. But all that depends on a fickle phenomenon known as the Loop Current—"

Click.

Really, I just need a few minutes of quiet. Let my head settle.

Pelicans remain much in evidence wherever the view embraces water. Terns. Laughing gulls. Salt marsh, rather pretty, at least for now. Occasionally a mullet jumps. I wonder whether they'll be here in these numbers next year.

I should hear what they're saying, so I turn on the radio again. ". . . They collected wind and current data from the last fifteen years, added an oil gusher, and ran it again and again to see where the oil would likely go. Oil ended up on the Atlantic coast of Florida more often than not. But the model could be wrong. The Loop Current usually pushes oil out of the Gulf and along the east coast of Florida, but it's not doing that right now. On the other hand, it looks like this could be a record-breaking year for tropical storms in the Gulf. This week's storm has delayed oil cleanup activities. What actually happens to the oil will depend on the weather. So keep watching those weather forecasts. . . ."

That's news you can't use. All that matters: Is the oil still gushing or stopped? Still gushing. That's all that matters.

A sign says, "Taking Oysters Beyond This Point Prohibited." The syntax raises the image of escorting oysters out into the bay. They need the word "from" between "oysters" and "beyond." At any rate, a few weeks ago, oyster rakers freckled the wide water. Now there are none.

Boom in channels. Boom under bridges. A long crescent boom arcs out from the shoreline, then simply ends. Protects nothing. The inescapable visual dominance of oil rigs. And this fiasco. I've never before seen a coast I hated.

Cedar Point Pier is on the north side of the bridge to Dauphin Island. Last time I was here the place was open. Now there's a hand-painted sign saying, "Closed."

A middle-aged woman named Jo is feeding the cats. Doesn't want "them to be victims, too." Takes me inside the store next to the empty piers. Shelves empty. "This is what's left." She's known me for two minutes, but she gives me a soft drink and a candy bar and won't let me pay.

On June 10 at 9:30 P.M., while several dozen people were fishing from the pier, police officers came onto the property to announce that the waters were closed to fishing and the place was "closed immediately." Not just the fishing. The whole business, bait shop, snacks and drinks, everything.

When the seventy-four-year-old proprietor came out to get an explanation, the six-foot-five officer knocked him to the ground to handcuff him, sending his face into a fence post on the way down and landing him in the hospital for two weeks. "Ambulance people could not believe the officer wouldn't uncuff him," Jo tells me. Word is that the officer had a prior problem controlling himself.

"When it's all over," she says, "it'll be bad for everyone."

While we wait to see when it'll be over, we pause for this word from the United Nations: a new report says that large modern corporations are "soulless" and threaten to become "cancerous" to society. The author of those comments, Pavan Sukhdev, is on sabbatical from Deutsche Bank and is working at the United Nations on a report called "The Economics of Ecosystems and Biodiversity." Reading between the lines, I infer that one goal of the report is to install an artificial heart in the treasure chest of multinational corporations. "We have created a soulless corporation that does not have any innate reason to be ethical about anything," he says. "The purpose of a corporation is to be selfish. That is law. So it's up to society and its leaders and thinkers to design the checks and balances that are needed to ensure that the corporation does not simply become cancerous." This doesn't come as news. But checks and balances—isn't that liberal code for regulations and taxes? In a famous report in 2006, British economist Nicholas Stern argued that the cost of tackling climate change would be 1 to 2 percent of the global economy, while the cost of doing nothing would be five to twenty times as great. Sukhdev says that failure to put a dollar value on nature's services—things like flood protection and crop pollination and carbon take-up by forests—is causing widespread destruction of whole ecosystems and of life on Earth.

"Arrogant and in denial." That's how a safety specialist who helped BP investigate its own refineries after the deadly 2005 explosion at its Texas City facility describes the company.

BP had grown into the world's second-largest oil company, behind only ExxonMobil. The company struck bold deals in shaky places like Angola and Azerbaijan, and drilled high in Alaska and deep in the Gulf of Mexico. It did "the tough stuff that others cannot or choose not to do," as BP CEO Tony Hayward once boasted.

When Hayward became BP's chief executive in 2007, he did away with fancy affectations, replacing art in the company's headquarters with photographs of BP service stations, platforms, and pipelines. A geologist by training, Hayward dispensed with the limousine used by his socially prominent predecessor. "BP makes its money by someone, somewhere, every day putting on boots, coveralls, a hard hat and glasses, and going out and turning valves," Hayward said in 2009. "And we'd sort of lost track of that." He vowed to make safety BP's "No. 1 priority."

But Hayward's predecessor, John Browne, had gone after the most expensive and potentially most lucrative ventures. And that record of risk translated to success. BP's share price more than doubled; its cash dividend tripled. Browne was knighted.

As Browne was reaching, he was also cutting. He outsourced. He fired tens of thousands of employees. Among them, many engineers. He kept rotating managers into new jobs with tough profit targets. The Texas City refinery, for example, had five managers in six years before the blast that killed fifteen people.

Built in 1934, that plant was poorly maintained even before BP acquired it. "We have never seen a site where the notion 'I could die today' was so real," a consulting firm hired to examine the plant wrote two months before the accident. The explosion was "caused by organizational and safety deficiencies at all levels of BP," the U.S. Chemical Safety Board concluded. The government ultimately found more than 300 safety violations. BP paid $21 million in fines, a record at the time. A year later, 267,000 gallons of oil leaked from BP's corroded pipes in Prudhoe Bay, Alaska. BP eventually paid more than $20 million in fines and damages.

Revisiting Texas City in 2009, inspectors from the U.S. Occupational Safety and Health Administration found more than 700 safety violations and proposed a record fine of $87.4 million. Most of the penalties, the agency said, resulted from BP's failure to live up to the previous settlement.

In March 2010, OSHA found 62 violations at BP's Ohio refinery. An OSHA administrator said, "BP told us they are very serious about safety, but they haven't translated their words into safe working procedures, and they have difficulty applying the lessons learned." On May 25, 2010, in Alaska—during the Gulf blowout—BP spilled about 200,000 gallons of oil. It was the Trans-Alaska Pipeline System's third-largest spill.

"In effect, it appears that BP repeatedly chose risky procedures in order to reduce costs and save time and made minimal efforts to contain the added risk," wrote Representatives Bart Stupak, Democrat of Michigan, and Henry A. Waxman, Democrat of California. "BP cut corner after corner to save a million dollars here and a few hours there," Waxman said. "And now the whole Gulf Coast is paying the price."

BP's executive overseeing the Gulf response, Bob Dudley, says it's unfair to blame cultural failings at BP for the string of accidents. "Everyone realized we had to operate safely and reliably, particularly in the U.S., to restore a reputation that was damaged by the accident at Texas City," he notes. "So I don't accept, and have not witnessed, this cutting of corners and the sacrifice of safety to drive results." He's about to become Tony Hayward's successor.

Dauphin Island, Alabama. Salt marshes on the north side. Egrets here look fine, miraculously immaculate. As do gulls. Adult pelicans—unlike those dingy birds near Grand Isle, Louisiana—retain the whiteness of their heads. Wherever they've been foraging, it's the safe place to be now.

On the main island, palmettos and cicadas. Full-on summer. In the entrance to a bed-and-breakfast hangs a big, framed prayer:

> We pray for your protection, Lord
> From the oil that is on the sea.
> We ask that you keep it far away
> And our island safe and free . . .

Standard selfishness directed skyward. How about asking the Lord to *stop* the blowout. And why would any God of mercy whose eye misses not the falling sparrow actually need to be asked?

Marion Laney, a part-time real estate agent here, says, "Every day something happens that makes you say—y'know—'What the fuck?' Stuff that doesn't make sense, at any level."

How's business? "I'm on the verge of bankruptcy in so many ways it's almost funny." Fourth of July weekend coming up. Last year one local company rented one hundred and eleven units for the holiday weekend. This year: ten. House values have fallen by half.

We park in the driveway of someone he knows. They're away. He wants photos of the four bucket loaders currently digging enormous amounts of sand from the north side of the island. The sand's being trucked to the ocean side, where other machinery patty-cakes it into ten-foot berms. Sand castles of woe. Also, unbelievably expensive.

They're digging a series of pits about one hundred yards long, fifty yards wide, twelve feet deep. Destroying whatever habitat was there. Scale of digging, scale of sand removal, scale of vegetation ruination: astonishing. Likelihood that this will deter oil in a hurricane: zero.

I'm guessing the idea is: attempt to replace sand dunes washed away by the last decade's major hurricanes. In other words, a subterfuge; the town is gouging BP while trying to fix an issue not related to oil.

Terns carrying fish circle the site; they seem to be looking for lost nestlings. Word is the permit was expedited. Loss of wetlands? What wetlands?

My other guess: BP doesn't care what excuse municipalities use to tangle themselves in BP's money. Money's how they gain control. Money's what they have. And all they have. Reasonable people might disagree with me. Unreasonable people surely would.

Also on the bay side, National Guard workers are replacing the natural shoreline with hard walls, armoring the shore against—what are they thinking? It all seems part of a "can-do" attitude that can't do the job of distinguishing whether the cure aids the ailment or kills the patient.

Anyway, when a car pulls into the driveway, Laney tensely breathes, "Car coming." He's afraid it's BP's Bully Police. How quickly the chill sets in. How easily. How thoroughly. When he recognizes the person, he says, "It's okay; I know them."

But it's okay either way, because all we're doing it taking photos from the deck of his friend's private house.

On the ocean side, a man is visiting a friend; he's brought along his family. His vacation home is in Gulf Shores, but Gulf Shores is closed. "And anyway, your kids go a couple hundred yards, then track tar all in your house and ruin everything you've got." He gestures with his chin. "These berms, they're useless."

He's part owner of an RV park. Recent acquisition. Bad timing. No tourists, no business, but it costs him twenty-three grand a month to pay his note. And in the midst of renovations, his contractors quit to work for Better Pay. "We have $3.2 million invested, and I just had to lay off all our help. In six months, we'll probably lose the place."

Right on the public beach where there's almost no vegetation left, I see a little puff of green near the end of a miles-long berm. Around this vegetation, a little string fence and a few small signs announce that a sea turtle's clutch of eggs lies incubating below the sand. All the other vegetation on the beach has been removed. If the nest hatches, the dark shape of the berm will look to the night-emerging baby turtles like land. Instead of marching to the sea, they'll march away from the berm, inland, and die.

At a pile of sand forty feet high, I climb to take photos. A guy in a truck from a company called Clean Harbors, from Albany, New York, gets out. I'm sure he's going to hassle me. Instead he says, "It's okay to park here. How can I help you today?" I say he already has, just by being such a breath of fresh air. I ask if his company is contracted by BP. "I really can't comment." Okay, something less threatening: I wonder out loud how much it's costing to make these miles-long berms. Just small talk. I'm not really asking him. He says, "I really can't talk about any of that. I could get fired for talking." The First Amendment protects corporations' free speech, but not the free speech of people who work for those corporations.

It's the Fourth of July. Two hundred and thirty four years ago, the United States won its independence from, guess who: the British.

Not so fast.

At the end of the road is a public park clearly marked "Open." It's noon when, on foot, I approach what looks like a little temporary trailer-style guard station at the entrance.

A young guy with a clipboard straightens up and walks quickly to the threshold of the park entrance. The power play is immediate. He's joined by an older guy, sixties, who takes over the interaction.

"I assume this park is open," I say, "since the sign says 'Open.' "

"No."

"Then why does the sign say 'Open'?"

"Because they haven't taken it down."

"Why would they take it down if it's a public park that's 'Open'?"

"It's closed."

"Why is it closed?"

"Because there are operations going on here connected to the oil spill. It's been taken over by the National Guard. The National Guard is operating down here."

"And they've closed the public park, even though it says it's 'Open'?"

"Yes. The city has closed it."

"Does that seem right to you?"

"I have no comment on that, sir. All I know is, it's closed to the public."

"And who do you work for?"

"I work for Response Force One security."

"What is Response Force One? I've never heard of that."

"A security company."

"And you're hired by who?"

"Response Force One."

"Yes, but who are they hired by?"

"BP," he says with a lift of his chin, as though those two letters are the big trump card of the whole Gulf region. A foreign-based corporation has hired American citizens to keep other Americans off public property clearly marked "Open" on our national holiday. These are not even real cops. They're what we used to call rent-a-cops, private security guards, the kind appropriate for guarding private property like office buildings and department stores. The kind who have no real legal authority. Local police or sheriffs, as I understand it, can grant authority to private security guards, but I can't check whether they've officially done so here, since, after all, it's a holiday. These guys, however, are not guarding the park or public property. They're guarding, well, I can't really see; looks like more booms and Porta Potties.

"So BP closed the public park?"

"No. The *town* has closed it. Because there are operations going on here."

The parking lot behind the guard is pretty empty, and the equipment is idle. It's a holiday, after all. "I don't really see any operations *going on* here," I say.

"Sir," he says, starting to lose his cool, "I'm tellin' ya—it's closed. Okay?"

"Okay, and I'm asking why."

"Because there are operations going on. It's a secure area."

"So, it's the Fourth of July, Independence Day, and—"

"Sir," he interrupts, now getting exasperated, "if you have any questions about the beach being closed, I'm gonna suggest that you contact somebody from the Town of Dauphin Island."

"Okay. Who can I contact?"

"Anybody in the city. Contact the mayor."

"Do you have their phone numbers?"

"They're not working today. It's a holiday."

"But do you have phone numbers, any contact information?"

"No, I don't." Now he's pretty fed up with me. Feeling's mutual, I'd estimate. The eye contact between us is turning hostile. I ask if I can take a picture of him and the younger guard at their booth and, not surprisingly, he says, "No."

"But you're on public property," I point out again. I assume he'd have a private right not to have his photo taken. Since he's saying he's acting in an official government capacity, however, I'm pretty sure I'd be within my rights taking a photo of an "official." But this is not a discussion on the fine points of the Constitution. We're miles and miles from the fine points. This, after all, is the Oil.

"*I said, NO!*" he yells. I see hatred in his eyes and he's starting to shake with rage. I guess he's not accustomed to being challenged. Most cars coming to the park just turn around upon seeing the guard booth. He usually doesn't even have to talk to anyone. His mere presence is enough to repel people. Everyone can see it's closed, never mind the sign. And obviously, I've come with an attitude about this. I hate all of it. I feel myself pointlessly returning his glare of rage.

He reaches for his radio in a threatening way, as if to warn, "I'm going to call Daddy." He's got his finger on the key.

I should make him call some real police, who work in tax-paid uniforms and drive a real cop car. But I presume they'll side with him and we'll all get angry. I turn and leave.

Later, I do find a sheriff and ask how private security guards can keep people off public property, especially where the sign very clearly says a park is "Open."

"I don't know," he says. "The park is closed; that's true. But right now there's a lot of screwy things on this island. I don't understand it all myself."

I guess you can't explain something that doesn't make sense.

Lunchtime. Shootin' the breeze over a beer with Marion Laney. He says, "There's a weird difference between what the government says to do and what this corporation does." He can't understand why, when the U.S. government tells BP to do something, BP seems to have the luxury of deciding whether to comply. "If this was a Venezuelan drilling company—or a *Cuban* drilling company," he chuckles, "I wonder what the government tone would be."

He's spoken to some of the fishermen whom BP is paying in the idiotically named Vessels of Opportunity program. Laney says, "Guys who'll talk when there's no camera around say they just putter around, not getting much done; say there's no real plan of attack. It's just, like, get in your boat, get out there, come back at quitting time."

He adds, "I hate to be a conspiracy theorist, but I think it's all just a big show. They're spending a lot of money. But either they're totally incompetent or there's some reason behind putting a lot of people out there and getting very little done."

Cooking up conspiracy theories here is easy as cereal and milk. Laney thinks BP wants the fishermen's mouths zipped—everybody just stay calm—and for a day's wages, they comply. But with bills to pay, what other option do they have? I think it's exactly that simple.

On docks where no one would think to wear life preservers just to walk to and from the boats they've run for decades, everyone wears life preservers. Former captains of their fate now march to rhythms not their own. Second childhoods in their own prime. Fishing boats lie stacked with white absorbent pads, like the diapers for whole communities suddenly severed from the normal progression of a life's memories.

At the marina, no one's in the store; they're not expecting customers. When a fisherman wearing a life preserver enters, we surprise each other. I'm not expecting "Abandon ship!" on land; I just want a few snacks. He calls a thirtyish woman, who takes her place behind the counter.

With a chuckle, she says, "We could be unemployed in the blink of an eye. Everyone's working for BP. That's all there is now. No one can think past next week." She says, "Everyone's worried in every way you can imagine."

I can't imagine. I get to go home.

Because it's a holiday weekend, Marion wants to see if he can get any shrimp. In the seafood shop over which he presides, Gary Skinner sweeps back his golden hair. He's a large man who describes his age as "fifty-nine, going on a hundred." He'd been a jeweler and watchmaker. "Got to be a dead trade," he says. "Cheap imports. Now we got cheap imported shrimp. That's why we had to open this shop up, to survive shrimpin'. Couldn't survive on wholesale prices."

We get a little tutorial on shrimp. A connoisseur, Skinner says, "The pink shrimp is probably the prettiest shrimp. It's beautiful. It's got this unique little black spot on the back of it; the shell is firm, it peels real good, it's real sweet. It's just a good shrimp. They're all good, but I think it's just the best."

He's had the shop for six years. He's been shrimping since 1975.

His two shrimp boats supply his shop. Except now they're working for BP.

"My big boat, last week they scooped up about forty barrels of oil. That's, like, sixty seconds' worth of what's comin' out that well."

He tells me, "At first, my wife said, 'There's gonna be a big slick comin' this way.' I said, 'Aw, they'll stop it in a coupla days.' When a week went by—. They started shuttin all the fishin' down. I mean, they *had* to, or you'd be getting oil all over your net, contaminatin' your whole catch.

"Right now, BP's actually payin' *more* than shrimpin' pays," Skinner says. He explains that the shop's business has grown 20 to 40 percent each year, including 20 percent in the 2009 recession. It's been a big success. "BP has figured in what our profit would have been this year, and what they've been giving me has been accurate," he reports. "They're payin' on time, so it looks like that's not gonna be a problem."

The problem: "Business has really fell off since this oil. This weekend—and this is a big holiday—it's down about 80 percent. Normally we have tourists, we have fresh shrimp off my boats. We'd sell forty or fifty ice chests full of shrimp for backyard parties.

"There's been a lot of sleepless nights," he acknowledges. "The money's one thing. Mainly it's been hard watching the business going down. What are we gonna do if they find out everything's contaminated, that it's killin' the nurseries inside, where shrimp grow. Or it's on the bottom offshore, where the shrimp spawn. It could be all over with. If they get it stopped, there might be some light at the end of the tunnel. Nobody knows, man. This is new for everybody."

His daughter-in-law works in the shop, with her babies calling and crying. His sons are running his boats. He comments, "My own two little boys started coming out on the boat when they were three years old. Now they're captains." He says he built the business to leave to his grandkids, "if they want to go fishin.' If they want to go to college and be a doctor, I support that, too." He adds, "People say, 'Oh, how can you keep doing this?' Well, we've had a good life. Money's not everything. We've had a lot of good times."

But we're not having good times today. This is the sorriest Fourth of July I've ever seen, and—because I like boats, I guess—it seems saddest at the boat ramp. Fishing is closed, but boating is allowed. Yet right now, the boat launch is empty when it should be packed.

Marion says, "Normally, there'd be dozens of boats here, jockeying for position, launching, hauling, standing off waiting to get in or out. It would be a madhouse. Normally there would have been hundreds of boats launched this morning, hundreds coming in at day's end."

There's not one single boat. No motorboats, not a single sail anywhere in view.

Dauphin Island has canceled its official fireworks. I guess no one is feeling sufficiently independent. I hear that BP offered ten grand toward fireworks but the town had the pride to decline.

Plenty of private fireworks, though. *Zip—pop!*

I get myself invited to a deck party. Lots of locals and a lot of great food on the grill. Nice view of the ocean, the sunset. Bottle rockets, Roman candles. *Zoom. Boom.* Firecrackers crackling by the pack—a regular good time after all, maybe.

We solemnly drink a shot of tequila to sundown. The whole scene is beachy enough to shake the fog of oil for a little while. Like, thirty seconds. It's all anyone is talking about. And here, where the blowout has turned things upside down, most people are thoroughly confused.

"How deep is the well?"

"I think it's eighteen thousand feet."

"It's seventeen *hundred* feet. The well itself starts five thousand feet from the surface."

"No, it's a lot more than seventeen hundred."

"I think it's seventeen thousand."

"I think that's right."

"Or it's seventeen thousand feet from the surface—"

"Maybe it's seventeen thousand from the surface to the bottom of the bore?"

"No, no; from the *floor* of the sea, it's seventeen thousand—. Uh, wait; you're right. Correct, correct."

"I don't understand these numbers; how can they do this drilling?"

"There's *nobody* in the Coast Guard who knows—I mean, the oil companies are *secretive.* You can't find out *anything* from the oil companies."

"That's—to me—the government should be keeping the transparency."

"Y'know, we put observers on fishing boats."

"We had observers in *Iraq.*"

"Observers do visit, but—"

"They should have observers on *all* these rigs. And they should *report* to the public what's going on."

"Instead, Minerals Management Service, literally in *bed* with them; what I hear."

"That whole department needs to be washed out. Start over."

"BP's hiring for a lotta different things. Look for oil, put out boom, check boom, move boom around—a lotta different stuff."

"What do they tell you *not* to do?"

"Can't pick up birds; gotta call them in."

"And they got, like, a gag order. They tell you anything you find is the property of BP."

"These past weeks have felt like months. You *know* it's gonna really hit. You're pretty sure, anyway. You don't know when. But with these onshore winds, you know it's coming. Don't quote me; I have a contract. Been to training. Got my yellow card. Got my credential card. Hopin' to get called."

A diesel mechanic says, "*All* the work's slowin' down. People don't use their boats, nothing breaks."

"Recently I went for three days' fishing in the blue water, about a hundred twenty-five miles from shore. These guys I was with are very knowledgeable about catching fish. We caught one dolphin and a barracuda. Normally at this time of year, we should've loaded the boat. We should've had lots of yellowfin tuna, wahoo, a *bunch* of dolphinfish, prob'ly hooked a blue marlin. It shoulda been *phenomenal.* This is a *very* good time of the year. We were shocked. We did not see oil—but the fish are *gone.*"

A man named Jack, retired after thirty years of working seafood safety for the Food and Drug Administration, says in frustration, "If only the guys from Transocean had said, 'Y'know, we've got a problem with one of the gauges not working right and we feel like we shouldn't remove the fluid from the well,' and the BP guy had said, 'Let's take a couple of extra days and do it right.' That's all they had to say. None of this would have happened. But they did the exact opposite. They forced this catastrophe on everybody."

"And the second wolf said, 'I can save you money by . . .' "

By the bottle rockets' red glare, I shake hands with a guy who says he was in Vietnam in 1968. I say, "That must have been unspeakable." He says it was. He says, "What kills me about this present atrocity with the oil is that when you've had an investment in this country like I have made, seeing the lack of mobilization is staggering. I can't believe this has been allowed to happen to the United States. I can't believe that the United States allowed this to happen to itself."

On July 7, the Thadmiral tells America via CNN.com that a relief well is "very close" to being completed, adding that he expects it to intercept its target in a mere month to five weeks.

I'm going to keep breathing.

The same CNN article informs us that Christian, Jewish, and Muslim clergy joined "in prayer and commitment to the communities most affected by the BP oil disaster," while "wives of current and former major league baseball players also fanned out across southern Louisiana to draw attention to the people and creatures affected by the disaster." Whatever it takes.

Louisiana's Governor Bobby Jindal, appearing to both lubricate his cake and eat it, complains that the oil has already ruined the seafood industry and depressed tourism, and now Obama's ban on deep exploratory drilling is costing thousands of Louisianans their jobs. When such wanting-it-both-ways rhetoric actually makes some sense—as this does—you know we're really in trouble. Stuck because we've built no options.

Bobby must therefore be pleased when, on July 8, a federal appeals court in the heart of New Orleans affirms a lower court's June 22 decision that there should be no drilling ban. Two of the judges on the appeals panel—both appointed by Ronald Reagan—had represented the oil and gas industries as private lawyers. Judge Jerry E. Smith's clients included ExxonMobil, ConocoPhillips, and Sunoco. Judge W. Eugene Davis represented various companies involved in offshore drilling. Blind justice.

Plaquemines Parish president Billy Nungesser, who's been outspoken throughout this ordeal, is apoplectic about the rule criminalizing getting within seventy feet of booms. Nungesser says the only way to maintain public confidence in the cleanup is to make it as transparent as possible.

I appreciate that sentiment, but I don't fully agree. In my opinion, if everyone really saw what a rope-a-dope circus this "cleanup" really is, they'd acquire scant reason for confidence.

But, actually helping address the confidence chasm, out of the ashes of the former Minerals Management Service, some fresh resolve. The former federal prosecutor who now heads the Obama administration's newly created Bureau of Ocean Energy Management, Regulation and Enforcement outlines for us his planned approach: "I'm not going to say you can't drill, but if people don't get the message that we are really stressing regulation and enforcement to an unprecedented degree they will have problems with me," he says. "There's a reason why we renamed the agency by putting regulation and enforcement in the name."

Just a few days after the appeals court affirmed the overturn of the administration's ban on exploratory drilling, Interior Secretary Ken Salazar issues a *new* moratorium. Rather than basing the ban on depth, which the courts called arbitrary, he bases it on what he should have based it on in the first place: "evidence that grows every day of the industry's inability, in the deep water, to contain a catastrophic blowout, respond to an oil spill and to operate safely." Salazar adds, "The industry must raise the bar on its deepwater safety." It sounds like what I want to hear: our government working to protect us.

Of course, the head of the American Petroleum Institute says the only thing he can say within the narrow confines of his job script: that the ban is "unnecessary and shortsighted," that it will "shut down a major part of the nation's energy lifeline," that it "threatens enormous harm to the nation." In other words, all the usual hype.

And, speaking of hype, a White House senior adviser this week calls the blowout the "greatest environmental catastrophe of all time." He adds that he's "reasonably confident" that all of the oil can be contained by the end of July.

If only *all* great catastrophes will be over in a week. Sad fact: this is *far* from the "greatest environmental catastrophe of all time." Actually using oil is the far greater environmental disaster; oil and coal are changing the world's climate, swelling the rising seas, and turning the oceans acidic. And in the Gulf itself, *getting* the oil has destroyed far more of the Mississippi River Delta's world-class wetlands than the blowout ever will.

Congressional briefing, Capitol Hill. Testimony. National Wildlife Federation president Larry Schweiger likens the response to inventing fire trucks and building a fire department for a house that's already in flames. He relates his recent experience aboard a boat in the Gulf amid "oil an inch-and-a-half thick in places" and "fumes overwhelming."

Another witness at the briefing says that a sixty-nine-year-old man with sixty-nine cents in the bank called into Larry King's Gulf telethon and pledged $10 on his credit card. Why doesn't such decency come in corporate proportions?

Courts have ruled that a corporation's first and foremost "responsibility" is maximizing profits for shareholders. But what if the courts had said instead—in consideration of the fact that a corporation's profitability benefits from military, copyright, and patent protections of the United States government—and by being ensconced in a great and technologically sophisticated country made stable by its laws and the peace its taxes buy—that corporations must budget 10 percent of profits (in the age-old tradition of tithing) to doing social good? What would the world look like then?

But *that* would be big government, interfering with *business.* Burdensome regulation! Meddling with the market. Happy to enjoy the meal and loath to wash the dishes, oil companies and other multinationals slurp their market-distorting taxpayer-funded subsidies like spaghetti. But this prevents new energy technologies and new companies from getting a toehold in a real market.

Sorry; the mind drifts.

Next witness. Brittin Eustice. Charter boat captain, about thirty. He takes people fishing for fun. Rather, took. Like the guy I talked to on Dauphin Island, Eustice says he recently fished an area about eighty miles south of Port Fourchon, Louisiana. Lookin' for the big boys: marlin, huge yellowfins. It was "very dead." They got one small blackfin and one dolphinfish in three days of fishing. "Usually that's what you catch in the first ten minutes," he says; usually you see lines of live floating weed and a lot of flyingfish. It was, he testifies, "extremely unusual." About the Oil and his life, he tells the assembled congressional folks, "No one knows what's the plan, what the effects will be. How can I plan anything?"

National Wildlife's Schweiger has the closing comment: "America needs clean energy . . ." He elaborates while my mind drifts again, because, of course, this is the most obvious message of the blowout. But before that locomotive can start moving, more fundamental things need to happen, things about money in politics.

⁂

On July 10, a Saturday, crews use undersea robots to yank six 52-pound bolts and remove the existing flange. They remove the ill-sealed cap installed five weeks ago. Oil resumes gushing. Next task: installing a 12-foot high, 15,000-pound piece of equipment that will eventually attach to the remaining hardware on the ocean floor and, on its other end, allow attachment of an 18-foot-high, 150,000-pound series of hydraulic seals that—engineers hope—will allow them to fully direct all the oil up to collection ships.

In a procedure taking several days, they install this new tighter-fitting cap, which features valves like a blowout preventer's. On a positive note: it is possible that the cap can also close the well like the blowout preventer was supposed to do, stopping the flow of oil altogether.

It appears that BP is actually getting serious about stopping the flow of oil. Not just collecting some of it. Not just marking time while waiting for relief wells that require more weeks of drilling.

Cleverly—how refreshing to be able to say that—this new cap has a perforated temporary pipe above the valves. Rather than trying to simply cap the enormous pressure, the perforations allow the pressurized oil to continue escaping through the open valves while workers fully tighten the cap into place.

By July 14, on the eighty-fourth day of the blowout, nearly 5 million barrels—something like 200 million gallons of oil—have spewed into the Gulf. And the well is still flowing unchecked. This blowout now clearly outranks Ixtoc I's 3.3-million-barrel blowout.

Engineers have been trying to determine if it's safe to close the cap's vents and let the pressure build. The worry is that pressurized oil could rupture another channel for itself right through the seafloor. If that happens, there'd be no way of controlling it. They're also afraid that the blowout might find its way into the relief wells, blowing them out, too. These fears cause BP to suspend digging the relief wells that we've been told for months are the "ultimate solution." Meanwhile, oil keeps gushing into the Gulf.

Finally they're ready to see if this cap is going to work. They begin closing a trio of ram valves. They monitor the pressure gauges. If the pressure holds, it means the cap is holding. If the pressure falls, it means the oil has found another route out.

New vocabulary term: "static kill." Definition: using a device that should already have been invented, built, tested, and warehoused for this purpose before the blowout ever happened.

The Thadmiral tells us that BP tells *him* that the test of the pressure will last six to forty-eight hours "or more, depending."

And as if our expectations could be lower, BP tells us to hold our applause, that such a cap has "never before been deployed at such depths" and that its ability to contain the oil and gas cannot be assured.

The pressure builds.

And so on July 15, a seismic shift: at about 3:00 in the afternoon, it looks like the new cap, with its valves shut, is holding the pressure. They've gotten control of the upward surge.

In other words, they've stopped the leak.

But no one celebrates. Everyone is holding their breath. No one knows whether the pressure will hold indefinitely. In the days to come, the pressure appears lower than expected, suggesting the possibility of leaks. Word of leaks near the well begins seeping out in the media. BP wants to leave the well capped and the valves shut. The Thadmiral wants BP to hook up the collection lines and let the oil go into waiting vessels at the surface. That would lessen the pressure, lowering the chances of a rupture through the nearby seafloor.

But it seems there are no leaks; they leave the cap closed.

This is how the blowout is over. It's the end of the beginning.

Eighty-six days. Two thousand and sixty-four hours.

Kill the well, slay the dragon, conquer the demon. But it's only liquid under miles and miles of the pressure of the Earth. Prick of a needle. That's all it ever was. The real fire-breathing dragon, the real dangerous demon, lurking on the surface all along, can be located in the mirror.

It's been the world's largest accidental release of oil into ocean waters. Twenty times the volume spewed by the *Exxon Valdez.* They'd told us 1,000 barrels a day; then 5,000; then 12,000 to 19,000; then upward from there. Now the best estimate: 56,000 to 68,000 barrels per day. From April 22 to June 3, oil had spurted from a jagged break in the riser pipe at about 56,000 barrels, or 2.4 million gallons, per day. After June 3, when robots cut the riser, oil spewed into the ocean even more freely for a time, around 68,000 barrels—approaching 3 million gallons—per day, or 2,000 gallons per minute.

The Gulf got an estimated 4.9 million barrels, about 206 million gallons: 99,700 gallons per hour, 1,662 gallons per minute, about 30 gallons per second. Roughly 25 gallons each and every time your heart sent a pulse of blood up your aorta, for nearly three months.

But the beat goes on and the usual pumping and burning continues; we in the United States burn over 20 million barrels of oil a day, about the same as Japan, China, India, Germany, and Russia—*combined.* Per person, Americans burn more fossil fuels and emit more carbon dioxide than anyone else; 4 percent of the world's population uses 30 percent of its nonrenewable energy.

Just as some of the hysteria was misplaced, now a false calm ensues. BP seems to be saying it's all finished; the oil is gone. All gone. That is also hype, a kind of information junk shot, pumping smoke and mirrors into the media stream to jam the flow of clarity.

BP's stock price rises.

LATE JULY

For a crisis begun so spectacularly, it's a murky, uncertain ending. We begin the next phase with an enormous amount of floating oil, and Gulf waters polluted by deep oil and dispersant. What now?

Of course, turtles, birds, fish, shrimp, crabs, shellfish, dolphins, and all their habitats have been affected. And creatures that range widely but funnel through the Gulf to migrate or to breed are of hemispheric importance. How many of these creatures, and what proportion of their populations, were damaged or spared? That, no one can say.

So far, the dead wildlife found and collected total about 500 sea turtles, 60 dolphins, and nearly 2,000 birds. It's not at all clear that oil killed most of them. But it's also not at all clear how many were killed by oil and never found. One dolphin's ribs were broken—a boat strike. But dolphins are usually nimble enough to avoid boats. A necropsy found what looked like a tar ball in its throat. Affected by oil, then hit by a boat; it's a reasonable conclusion. But not certain.

Every oiled carcass found may suggest ten to one hundred undetected deaths. Birders are reporting the disappearance of black skimmers, pelicans, royal terns, least terns, Sandwich terns, laughing gulls, and reddish egrets from several Louisiana islands, including Raccoon Island, Cat Island, Queen Bess Island, and East Grand Terre Island.

But it's also true that there remain along Gulf shores and marshes many clean-looking pelicans, gleaming gulls, and egrets with no regrets. A silver lining for animals with gills is that they got a season's break from fishing, which typically kills millions of adult fish, shrimp, and crabs, probably far more than this blowout did.

Some say the floating oil is blocking light needed by the plankton that are the base of the whole food web. Some say oil will settle on the bottom and continue washing up for years. Some say microbes will eat it. Some say those microbes will also rob regional waters of oxygen, killing nearly everything for miles. Some say no one knows what its effects will be, or how long they'll last.

As a naturalist, I find that the effects on wildlife remain hard to grasp.

People, though, have taken deep and immediate hits. Closures have meant an end to fishing, the cessation of a way of life and of the way thousands of people understand who they are. No one knows whether the seafood and tourism and fisheries will be clean and healthy again next year or in a decade.

So even with the leak stopped, there's hardly a sigh of relief.

"Just because they kill the well doesn't mean our troubles go away," says the crab dealer in Yscloskey, Louisiana. Another fisherman says, "As soon as BP gets this oil out of sight, they'll get it out of mind, and we'll be left to deal with it alone."

"We're not going anywhere" is the reassurance offered by NOAA chief Dr. Lubchenco. But official reassurances do little to dispel such deep anxiety. Many residents don't know who to believe, or what to believe. Many simply believe that the oil isn't going away anytime soon, whatever anyone says.

But they do believe that help is going away. During a public forum, Louisiana fishermen hear an Alaskan seafood spokesman describe how after the 1989 *Exxon Valdez* accident, the perception was that oil affected all seafood from the state. "It took ten-plus years to get out of that hole," he says. "We put in a request directly to the oil companies to fund a marketing effort. They gave us zero, absolutely nothing."

An estimated 45,000 people and 6,000 vessels and aircraft have by now been involved in the response. Now even the *retreat* of oil manages to become bad news for fishermen. For many fishermen idled by the oil, responding to the oil has become their livelihood. As the crude ebbs, many face being left with nothing to do but wait for the government to reopen fishing areas, and for consumers to show confidence in their seafood. And there are no guarantees that they will be able to resume fishing soon.

"This whole area is gonna die," bemoans a fifth-generation fisherwoman in Buras, Louisiana. "Down here, we have oil and we have fishing," she says. "We are water people. Everything we do involves the sea, and the spill has taken it all away from us. I've been contemplating suicide to the point of making myself a hangman's noose, honest to God. Then I decided that's not going to do anything, apart from shut me up."

The manager of a BP decontamination site for cleanup workers says he gets calls "every day" about people who want to commit suicide.

I can't help noticing it's only the victims who want to kill themselves.

On July 19, while a tanker is offloading oil near a Chinese port, two oil pipelines explode, sending flames 60 feet into the air and an estimated 11,000 barrels of oil into the Yellow Sea. Local media reports a 70-square-mile slick, which workers try to break up with chemical dispersants and with oil-eating bacteria.

On July 21, a federal judge stops companies from developing oil and gas wells on billions of dollars' of leases off Alaska's northwest coast. He says the federal government sold drilling rights without following environmental law. Alaska Native groups and environmentalists insist that no one could ever clean up an oil spill in Arctic waters, especially if there's sea ice. The Arctic is much more remote than the Gulf of Mexico. It's far from harbors and airports. The nearest Coast Guard base lies many horizons away.

"That spill in the Gulf, it could have been our ocean," says Daisy Sharp, mayor of Point Hope, an Inupiat Eskimo community of seven hundred people on the Chukchi Sea. "It's sad to say, but in a way I'm glad it happened. Maybe now people will take a closer look at offshore oil drilling."

Caroline Cannon, Native president of the village, says the decision brought tears of joy: "The world has heard us, in a sense. We're not on the corner of the back page. We exist and we count."

Well, Ms. Cannon, you guys exist as far as I'm concerned, but oil greases Alaska. Alaska's governor and other politicians love oil, because the petroleum industry pumps more than 90 percent of the state's revenue into Alaska's budget. And that judge didn't void the leases, but merely ruled that the federal government must analyze the environmental impact of development. For the oil companies, it's not over.

For the people, it might be. "Our ancestors had seen the hunger for oil when the Yankee whalers came in the nineteenth century," says village council vice president Steve Oomittuk. "They came for the whale oil and wiped out the whales." Whole villages that had relied on whales for food disappeared, Oomittuk says. He adds, "Then six years ago we saw the hunger for oil coming back. We started to think, 'This time *we* will go extinct.'"

But a whale might think, "Thank God for petroleum." Without it, as Oomittuk implies, we would have destroyed the whales. And thank heaven even more for the power of the sun, the wind, the tides, the heat in the heart of the Earth, the oils in the algae that feed the whole sea, these eternal energies that drive our world and all its life. Without them, we might have overheated the planet and acidified the ocean. But I'm getting way ahead of the story; we're not there yet. Not even close. As the whalers were stuck in their remorseless havoc, so we have stuck ourselves, with oil.

As we burn the easy oil and tap the deep oil to the limits of technology, sources that hadn't been worth it are getting attention. Enter: "oil-sands," "tar sands," "shale oil," and other relatively meager sources that are now worth money. Of Canada's northern Alberta—a province I remember as beautiful when it was younger—I read that as the Athabasca River and several of its tributaries flow past facilities gouging at oilsands (one ugly word), heavy metal neurotoxins like lead and mercury are entering the water at levels hazardous to fish. Add to them cadmium, copper, nickel, silver, and seven other metals considered priority pollutants by the U.S. Environmental Protection Agency. First, workers bulldoze the trees and strip the soil. "As soon as there was over 25 percent watershed disturbance we had big increases in all of the contaminants that we measured," one scientist tells us. What's he mean by "big increases"? In places, cadmium levels ranged between thirty and two hundred times over the guideline set by the Canadian Council of Ministers of the Environment to protect marine ecosystems. Silver levels were thirteen times higher than recommended at one site, and copper, lead, mercury, nickel, and zinc were five times the suggested limit.

And then there's "fracking," wherein engineers actually pump fluids at high pressure to shatter rock formations up to several miles underground to get the gas they contain. This practice is increasing because the resources are getting depleted. It's a long way from tapping surface seeps. In the East, the big target is the Marcellus Shale, which underlies states from Virginia to New York and into Canada. Its recoverable gas has been estimated at 49 trillion cubic feet, about two years' total U.S. consumption, worth about $1 trillion.

If you've already heard of fracking, here's a new vocabulary word: "proppant." In a sentence: "Proppant, such as grains of sand of a particular size, is mixed with the treatment fluid to keep the fracture open when the treatment is complete." This is all injected deep into the Earth. In shales they use oil-based drilling fluids when they're getting down to the reservoir, and then the fracturing fluids pumped down contain a multitude of sins—proppants, gels, friction reducers, breakers, cross-linkers, and surfactants similar to those in cosmetics and household cleaning products. These additives are selected to improve the "stimulation operation" and the productivity of the well. The fluids are 99.5 percent water, but it's the other .5 percent that matters. The

New York State environmental impact statement for drilling to frack the Marcellus Shale has six pages of tables listing the many components of fracturing fluids.

You can begin to imagine the above-ground mess and risk of all this fluid. Then there is the little issue of drinking water. Experts say it's not a problem. The New York State environmental impact statement reads, "Regulatory officials from 15 states have recently testified that groundwater contamination from the hydraulic fracturing procedure is not known to have occurred despite the procedure's widespread use in many wells over several decades."

The Environmental Impact Statement says that there is a vertical separation between the base of any aquifer in New York (850 feet) and the target shales (below 1,000 feet, although it also, confusingly, gives this depth as above 2,000 feet). It says the rock between the target shales and the aquifers is impermeable, so it should be an effective migration barrier.

Go ahead and take a big sigh of relief.

Now get worried again. The big case everybody cites is Pavillion, Wyoming, where gas fracking has been happening for a while. After several years of complaints by residents, the U.S. Environmental Protection Agency sampled nineteen drinking-water wells and, in August 2010, confirmed in eleven the presence of 2-butoxyethanol phosphate (probably from frac fluids), plus adamantane compounds (definitely from fracking) in four wells, and methane in seven. This is the first confirmed case of frac fluids getting into groundwater. It's possible that the fluid contaminants got into the wells—into people's drinking water—from surface spills; but the methane is conclusively from the underground reservoir. The final EPA Pavillion report is recent and is the smoking gun for opponents of fracking.

And now a dash of good news, a step toward the techno prep that was needed in the Gulf all along: ExxonMobil, Chevron, ConocoPhillips, and Royal Dutch Shell announce plans to voluntarily contribute $250 million each to build what they should have had warehoused already: modular containment equipment that would be on standby, capable of capping a blow-away well or siphoning and containing up to 100,000 barrels of oil daily in the next leak or spill. (BP is not included initially but will join in September 2010.)

Ulterior motive: the companies want Obama to lift his ban on deep exploratory drilling. They also realize that the gushing hemorrhage of bad publicity is detrimental to their stranglehold on what should be a national discussion about our energy future.

"It's doubtful we will ever use it, but this is a risk-management gap we need to fill in order for the government and the public to be confident to allow us to get back to work," the oil spokesman says. It could be sixteen months before the system is completed, tested, and ready to be used. In a blowout, it could still take weeks to stop the flow. Drawings of the proposed system show a cap and a series of undersea pipes and valves and a piece of equipment that would pump dispersant. Lines would be hooked up to vessels on the surface.

If it doesn't work, there's still no good plan for capturing oil.

The oil companies say this initiative is the product of *four weeks* of intensive efforts involving forty engineers from the four companies. And, of course, they want to deflate the momentum for a series of congressional bills that aim to bring more safety, more oversight, and more unwelcome lawmaker meddling.

One bill they don't like, for instance, would force companies to drill a relief well alongside any new exploration well. Oil executives argue that this would double the risk because a relief well would be just as likely to blow out. They don't mention that this would be true only if they drilled the relief well all the way into the hydrocarbon zone—or that they don't like the fact that it would cost them a bundle more.

But what about that slippery and elusive bigger picture?

In 1969, Senator Gaylord Nelson was so moved by seeing the devastation following the oil blowout off Santa Barbara that he called for a national day to discuss the environment. After the resulting establishment of Earth Day the following year, Republican President Richard Nixon created the Environmental Protection Agency and signed the National Environmental Policy Act and the Clean Air Act into law. Congress followed that burst of legislation with the Clean Water Act, the Endangered Species Act, the Coastal Zone Management Act, the Toxic Substances Control Act, and other giant strides of high-minded lawmaking that invented environmental protection and put the United States ahead of any nation on Earth.

Though the Gulf eruption far exceeds Santa Barbara's ten-day, 100,000-barrel blowout, when President Barack Obama pushes for clean energy, Republicans accuse him of trying to exploit the Gulf tragedy for political gain. Liberals fault him for failing to specifically push for a cap on carbon dioxide emissions.

Santa Barbara's leak originated much closer to shore, covering miles of California beaches with thick crude oil. Pictures of dead seals, dolphins, and thousands of birds horrified the country. I myself had never imagined a bird coated in oil, and in my mind I still see vividly the television image that shocked me. Santa Barbara's psychological effect was huge. And the country was more cohesive. We hadn't yet had thirty years of anti-government propaganda and deregulatory chaos, so we set our government to work passing laws to address the problems.

That was then, this is now. South Carolina republican senator Lindsey Graham worked on a climate-change bill for months before pronouncing it hopeless. And now, in July's last week, legislation to reduce climate-warming greenhouse gases from fossil fuels implodes in the Senate, derailing the year's top environmental goal. Not even this disastrous blowout could create a national consensus to move America off dirty fossil fuels and into clean, eternal energy.

Lois Capps, now a Democratic congresswoman representing Santa Barbara and the central coast, was a young stay-at-home mother in 1969. She reminds us that Gaylord Nelson's speech came eight months after the January 1969 blowout. The resulting Earth Day didn't happen until April 1970. Not until 1981 did Congress impose a ban on offshore drilling along most of the nation's coastal waters, an action rooted in the memories of the Santa Barbara spill, more than a decade earlier. (That congressional moratorium endured for a quarter century; Congress lifted it in 2008.) "It doesn't happen overnight," Capps says.

Agreed, but it's now been 40 years—14,600 nights—and we still don't have a clean, eternal-energy economy. My friend Sarah Chasis, who started working at the Natural Resources Defense Council while I was still a college undergrad, says, "I think we're going to see a really significant response to what happened in the Gulf play out over time. I think it's going to affect people's thinking and the way they approach issues for a long time."

I'd like to think so. An Associated Press poll in June 2010 found that 72 percent of Americans rate the environment as "extremely important" or "very important," up from 64 percent in May and 59 percent in April. But that's the problem. We care when it's headlined, but it's toast when it's redlined.

We're better than we were in 1969 on important issues like civil rights, women's equality, even on the environment. But that's mainly because of what was accomplished *then.* Since then, we've increasingly given our economy and our government to the kinds of people who care more about themselves than they do about our economy or our government. They've stagnated our energy policy, shipped our manufacturing jobs overseas, and racked up enormous national debt. Why? What was accomplished for America? Mainly, very rich people became extremely rich. Which would be fine, but the widening gap in the middle is very harmful to most Americans. And that gap is where moderation would lie, and from where a sane, enduring, future-forward energy policy would emerge. Thoughtful planning doesn't come from people who think only of themselves and only of today, traits shared by both the needy and the greedy.

In the deep earth's wings, the relief wells continue grinding along toward their seven-inch target, the intention still being to "kill" the well at its bottom.

On July 22, the National Oceanic and Atmospheric Administration reopens commercial and recreational fishing in 26,388 square miles of Gulf waters, or about one-third of the 83,927 square miles of Gulf of Mexico federal waters that had been closed to commercial and recreational fishing—which itself had been over a third of all of the Gulf of Mexico federal waters.

Fishing means seafood means consumer confidence. So to rebuild consumer confidence, the government hires people to *sniff* seafood for oil. We're told the sniffers' identities will be kept secret to prevent harassment. That seems ludicrous, but then again, this is the blowout. So only the nose knows.

Inanity 101: A professor shows us how to sniff seafood: "Take small bunny sniffs," she says. No, really.

After the first clean samples, the professor demonstrates with a bowl of shrimp intentionally tainted with a small amount of ammonia. One small bunny sniff, and the verdict?

"It's pungent and putrid. It smells bad. Eww."

Well, *you put ammonia* in it. You didn't even put actual petroleum in it.

"I didn't want to use the actual oil with some of the volatile hydrocarbons that are in there just, you know, for safety purposes."

Oh, *puleeez!* We who run our lives with gasoline and heat our homes with oil and methane and propane will have our safety compromised by a bunny whiff of *oil?* We *live* in the stuff.

The whole idea of sniffing as a test of seafood safety seems iffy. But a deputy from NOAA seeks to reassure us by saying, "If you think about your ability to detect something in your refrigerator, if it has an off odor, you can detect it at very, very low levels."

So scientific. Don't you feel confident now?

We're told that if the sniffers *don't* find oil, the samples they've sniffed will get chemically analyzed. I hope so, because I have zero confidence in sniffing as a way to ensure the safety of massive quantities of seafood.

And get this: we're told that the sniffers will "detect whether a sample is normal by testing it against a baseline of seafood caught before the spill." In other words, seafood that's been frozen for more than three months will be sniff-compared with fresh-caught seafood, to see which smells better.

We're told, "The experts can detect contamination of one part per million. Newer, less experienced state screeners, up to ten parts per million—that's a single drop in a gallon of water." We're told, "You have known experts who are in the room who, in fact, can help direct the trainees towards this sort of smell you should be getting. You might get it in your tongue. You might get it in the back of your nasal cavity. You might feel it right here."

I feel like I'm back in the Middle Ages. And while we're sniffing, I'd like to know why we aren't using dogs. Do we have agency experts sniffing for dope and bombs in airports? Do expert hunters chase rabbits with their expert noses to the ground? C'mon.

We're told, "NOAA officials are on high alert to prevent any seafood containing oil from being caught and sold." The professor says, "Believe me, I don't believe someone would go ahead and attempt to eat something that is tainted, because it is very aromatic and it's quite unpleasant."

Really? Is that all you've got? Sorry; I have a lot of respect for professors, but I don't believe her. If tainted food always smelled bad, people wouldn't get sick from eating tainted food.

We're told that in addition to using sniffers, the agency sends hundreds of samples to a lab in Seattle for chemical analysis. So far, just one sample came back tainted. *That*, I believe. I'd reserve small bunny sniffs for detecting, say, carrots.

Despite splotches of brown crude that continue washing up here and there, Louisiana plans an early August opening of state waters east of the Mississippi to catching fish like redfish, mullet, and speckled trout. Shrimping season will begin in mid-August. But oysters and blue crabs will remain off-limits. "I probably would put oysters at the top of the concern list and I don't think there's a close second," says Dauphin Island Sea Lab director George Crozier.

What about seafood in general? The oil contaminants of most health concern, polycyclic aromatic hydrocarbons, or PAHs (which can cause cancer), also show up in other everyday foods, such as grilled meat. A NOAA spokesman says the levels in Gulf seafood "are pretty typical of what we see in other areas." That's because they're common in oil, vehicle exhaust, food grown in polluted soil, tobacco smoke, wood smoke, and meat cooked at high temperatures. NOAA found that Alaskan villagers' smoked salmon contained far more PAHs than shellfish tainted by the *Exxon Valdez* spill. Live fish metabolize PAHs rather than store them.

For instance, the highest level of the PAH naphthalene found in fish from recently reopened Florida Panhandle waters was 1.3 parts per billion, well below the federally considered safe limit of 3.3 parts per billion. Federal regulators say they're sure the fish will be safe. They say that of the more than 3,500 samples taken from the Gulf during the spill, none contained enough oil or dispersant to be harmful to people.

In fact, regulators did not turn up a single piece of seafood that was unsafe to eat—even at the height of the eruption. Unlike certain contaminants, such as mercury, which accumulates in fish, fish quickly metabolize the oil's most common cancer-causing compounds. Crabs and oysters metabolize these chemicals more slowly. So far, shellfish testing is just beginning, and many shellfish areas remain closed.

It's not appetizing to think about eating a fish that's been neutralizing cancer-causing chemicals, no matter how promptly and efficiently. But I'm confident that an occasional meal, or a bit more, would be safe. Safety is a relative term, of course. I drive a car—the most dangerous thing most of us do routinely. We live in a world of hazards, including some pretty awful artificial chemicals and food containing God knows what. On the other hand, we have higher average life expectancies than ever in history, especially if you subtract people who eat too much and who smoke.

So if the feds say the seafood is safe, I would be willing to put a little initial squeamishness aside. Others have their own analysis. "It tastes like fish," a fisherman says. "I'm not dead yet. If the government tells me it's unsafe, I won't eat it. But I'm not going to worry about it. They're the experts, I'm just a fisherman." A tourist finds a different source of confidence. "Restaurants aren't going to sell the shrimp if they're bad," she says. "These are decent American people down here. They're not going to lie about it."

You'd think anyone in the fish biz would be delighted at the openings, but some think it absurd to open fishing grounds so soon. Some fishermen and their families worry that if any seafood diners wind up with a tainted plate, it will be a knockout blow. "I wouldn't feed that to my children without it being tested—properly tested, not these 'Everything's okay' tests," says a shrimp boat owner from Barataria, Louisiana. She and her husband are sufficiently skeptical of the government's assurances that they will not go shrimping when the season opens, at least not initially. Gotta respect them for that.

For the owner of a Venice shrimp and crab processing dock, it's the not knowing. "Will there even be a market for Louisiana seafood?" he wonders. "What is the impact on crabs and shrimp over the long haul? It's impossible to know."

In the short haul, part of the answer is possible to know. Keath Ladner's Gulf Shores Sea Products has a steady customer who buys two million pounds of shrimp a year. They've canceled their entire order. "The sentiment in the country is that the seafood in the Gulf is tainted," Ladner says. "People are scared of it right now."

Dawn Nunez's family is also in the shrimp business, but she dismisses the reopenings as premature, "nothing but a PR move. It's going to take *years* to know what damage they've done," she denounces. "It's just killed us all."

July 28 is day 100. Louisiana gets a new small oil spill just after midnight when a barge being pulled by a tugboat crashes into an inactive well in Barataria Bay, sending a mist of oil and gas a hundred feet into the air, creating a mile-long slick.

Two days later, a ruptured pipeline gushes as much as a million gallons of oil into Michigan's Kalamazoo River, sending federal and state government officials scrambling to stop the oil from reaching the Great Lakes.

On July 30 Dr. Caz Taylor of Tulane University copies me on an e-mail saying, "We have been seeing droplets of what could be oil or dispersant (or both) in crab larvae, from Pensacola, FL, all the way down into Galveston, TX. We haven't yet confirmed what these droplets are but if they are oil-spill related, then we have been seeing effects of the oil spill in places where oil is not visible for a while now. So the (welcome) news that the surface oil is receding does not greatly change my perception of the magnitude of the effects of the spill. There is still a lot of oil out there. If the oil and/or dispersant has entered the food web then the effects will be felt throughout the Gulf although they may take months, or longer, to manifest themselves."

Here is what most people read in something like that: "We have been seeing droplets of oil and dispersant (both) in crab larvae from Florida to Texas. So we have been seeing effects of the oil spill even where oil is not visible. The (dubious) news that the surface oil is receding does not greatly change the magnitude of the effects. There is still a lot of oil out there. It has entered the food web and the effects will be felt throughout the Gulf for months, or longer."

Here's what a scientist reads: "We have been seeing droplets of what could be oil or dispersant in crab larvae, but we haven't yet confirmed what these droplets are. If the oil and/or dispersant has entered the food web, then the effects may take months, or longer, to manifest themselves. That's a big 'if.' "

Here's what the media actually writes and what most everyone reads: "University scientists have spotted the first indications oil is entering the Gulf seafood chain—in crab larvae—and one expert warns the effect on fisheries could last 'years, probably not a matter of months' and affect many species."

Meanwhile, engineers are in the process of pumping drilling fluid that, under miles of its own weight and pressure, can push the petroleum and methane back into its genie bottle three more miles to the bottom of the well. This not only holds the pressure in, it neutralizes it.

Success arrives quietly on August 5, after they've followed the drilling fluid with tons of cement, which seals off the well's walls and once again isolates the oil and gas back in its cave, where it had slept for millions of years, away from the living world.

At the end of July, BP posts a quarterly loss of $16.9 billion. BP must now coordinate the drudgery of picking up 20 million feet (3,800 miles) of boom.

With the spigot capped, talk of relief wells ebbs. Two relief wells were supposed to be completed by early August, which, in May, seemed unendurably far into the future. Now the latest tropical storm, which has again forced evacuation of offshore crews, has everything further delayed.

PART THREE

AFTERMATH

DOG DAYS

When I was a kid, we were told that oil formed from dead dinosaurs. The idea is so easy to visualize, it had persisted since the early twentieth century. But back in the 1930s, a German chemist, Alfred E. Treibs, discovered that oil harbored the fossil remains of chlorophyll; the source appeared to be the planktonic algae of ancient seas, blizzards of microscopic sea life gently falling into the depths over the ages. Covered with sediments, cooked by the geothermal energy of the planet's hot heart, dead microscopic algae became oil.

Some of the waters that made the planet's oil still exist, like the Gulf of Mexico, which has long received the flows and nourishment of big rivers draining the continent, as the Mississippi does today. Meanwhile, the seas that produced other massive oil fields, such as the Middle East's, are gone.

For over half a billion years, incompletely decayed plant and animal remains from countless quintillions of tiny organisms buried under layers of rock have been percolating in the pressures and boiling heat of the Earth. Since the Paleozoic era, roughly 540 to 245 million years ago, organic material has been slowly moving to more porous layers of sandstone and siltstone, accumulating there and pooling where it has become trapped by impermeable rock and salt layers. A typical petroleum deposit includes oil, natural gas, and salt water. Crude oil is a mixture of different hydrocarbons with different boiling points that facilitate their separation into materials like asphalt, heavy and light oils, kerosene, light and heavy naphtha (from which gasoline is made), gas, and other components. It gets further refined and blended with other products to make fuels, solvents, paints, plastics, synthetic rubber, soaps, cleaners, waxes and gels, medicines, explosives, and fertilizers.

Petroleum seeps to the surface in many places, most famously the La Brea tar pits in Los Angeles, which for tens of thousands of years captured mastodons, saber-toothed cats, giant ground sloths, lions, and wolves, and today continues to claim pigeons and squirrels. In the early days of Spanish California, settlers used La Brea's asphalt for roofing.

Though the world has relied on petroleum as a major industrial fuel for only a little over a century, people have been using petroleum for over six thousand years. The Sumerians, among others, mined shallow asphalt for caulking boats and for export to Egypt, where it was used to set mosaics to adorn the coffins of great kings and queens. In about 330 B.C., Alexander the Great was impressed by the sight of a continuous flame issuing from the earth near Kirkuk, in what is now Iraq; it was probably a natural gas seep set ablaze. People being what we are, the potential for petroleum-based weapons was recognized early on. Arabs used petroleum to create flaming arrows used during the siege of Athens in 480 B.C. The Chinese, around A.D. 200, used pulleys and muscle labor to pump oil from the ground, send it through bamboo pipes, and collect it for fuel. Not until the 1800s did the West catch up to this level of oil drilling. The Byzantines in the seventh and eighth centuries hurled pots filled with oil ignited by gunpowder and fuses against Muslims. Similar bombs used at close range nearly destroyed the fleet of Arab ships attacking Constantinople in 673. Bukhara fell in 1220 when Genghis Khan hurled pots of naphtha at the city gates, where they burst into flame.

During the Renaissance, oil and asphalt from shallow pools discovered in the Far East found their way to Europe, and traders soon established routes to the West. By the 1600s, petroleum was lighting streets in Italy and Prague. By the end of the Napoleonic Wars, asphalt was being used extensively to build roads.

In the New World, natives in what is now Venezuela used petroleum to caulk boats and baskets and for lighting and medicines. In 1539, a barrel of Venezuelan oil was sent by ship to Spain to soothe the gout of Emperor Charles V. In North America, certain natives used oil in rituals and for making paints.

The modern commercial petroleum era began in 1820, when a lead pipe was used to bring gas from a natural seep near Fredonia, New York, to nearby consumers and a local hotel. In 1852, Polish farmers in Pennsylvania asked a local pharmacist to distill oil from a local seep. They were hoping to make vodka, but the result was undrinkable. It burned, though, and so they invented kerosene. The invention of the kerosene lamp two years later created a mass market for commercial kerosene, and soon towns everywhere were glowing with the light of petroleum. In 1858, Colonel Edwin Drake pounded a well sixty-nine feet into the ground near Titusville, Pennsylvania. It produced a continuous flow of oil, and within a short time kerosene replaced whale oil for lighting lamps. In the 1880s, the first oil tanker began carrying oil across the ocean. By the 1970s, the seas began bearing thousand-foot-long supertankers capable of containing 800,000 tons of oil. Today oil accounts for over half the tonnage of all sea cargoes.

Petroleum didn't really get big until the internal combustion engines of the twentieth century and the autos, tractors, trucks, and, eventually, aircraft they powered. Middle East oil ramped up rapidly in the late 1940s. Oil was discovered in Iran in 1908, in Iraq in 1927, and in Saudi Arabia in 1938. By the 1970s, the oil fields of the Middle East were producing about half the world's oil. Between 1950 and 1970, annual world oil production surged from 500 million to 3 billion tons. Today there are more miles of petroleum pipelines than of railroads.

Saudi Arabia is now by far the world's largest oil producer. (The United States is the Saudi Arabia of coal.) The Organization of Petroleum Exporting Countries (OPEC) has the greatest oil reserves. Its member countries include some notable enemies of the United States, and some friends the likes of which make enemies unnecessary.

In many parts of the world, depletion of land-based oil has forced drillers to the continental margins and beyond. In 1947, the first in-water oil well began operating from a wooden platform in sixteen feet of water off Louisiana.

Today one of the most innovative and promising ideas for a new liquid fuel to replace petroleum involves extracting energy from the oil in genetically engineered algae. Algae make oil that can be converted to biodiesel; they can be used to make ethanol; they can be converted to biogas; and they take carbon dioxide out of the air. It may seem surprising that algae create long-chain hydrocarbons resembling petroleum crude oil, but it shouldn't. Energy from the oil in algae, it turns out, is pretty much what we've been using. Algae is, though, vastly better than petroleum in terms of risks and consequences. Unlike such other fuel sources as corn, soybeans, and sugarcane, algae do not compete with our food supply or, like palm oil, cause farmers to cut down tropical forests. And because algae absorb carbon dioxide—the main pollutant from burning fossil fuels—their use as fuel could actually help reduce global warming. Algae naturally produce about half of the oxygen we breathe. And they can produce a much larger harvest per acre than other energy crops.

ExxonMobil said in 2009 that it would invest $600 million over the next few years to produce algae fuels comparable to fuels refined from conventional crude oil. So the tiny organisms that produce petroleum may be the liquid fuel of the future—without us having to wait, say, 300 million years.

Sounds good. Why aren't we rushing to do this? Why isn't this a national priority?

For a very short course on oil's influence in Louisiana politics, meet Professor Oliver Houck. In a Tulane University Law School office with well-stocked bookshelves and a small aquarium, the floor piled with documents organized for a new book he's writing, Houck sits back casually in his swivel chair.

He sees Louisiana as a petro-state, its petro-dictators propped up by oil money spread across parties. No candidate can really stand up against oil and gas. Louisiana, in too deep, is stuck.

Houck notes, "There's a kind of toilet mentality in states where resources are abundant. There's no ethic of conserving. In a timber state, the feeling is there's nothing wrong with clear-cutting. In Wyoming, it's a crime to say something bad about coal. In Texas, you can't say something about beef. Florida's never been oil-dependent; it's dependent on white-sand beaches. Oil impregnated Louisiana politics a long time ago."

But we are a union of fifty states. Nationwide, might this blowout eventually change the mood on energy?

"In the people, yes. In Congress, no. And there's no connection. Nationally there's a feeling of being fed up with oil; that the oil companies can't be trusted. I think the people would strongly support an energy bill that would go far beyond anything now in Congress. But they're not passionate about it. So members of Congress don't fear getting tossed out by voters based on their record on energy. And in Louisiana we have two of the most sorry-ass senators in Congress. They're hacks. But they're hacks with enormous power, because whatever party is in the majority is there by just a razor-thin margin. So suddenly they're swing hacks with enormous influence."

You will not be surprised to learn that Louisiana's congressional representatives have tried to get rid of Houck and this law clinic. "The only reason I'm not gone," Houck explains, "is that I'm at a private university."

On TV, Congressman Ed Markey is complaining that the oil companies "basically owned and operated their regulator but [the blowout] will catalyze Congress to create legislation to end this. We cannot continue to allow their cozy complacency between regulator and regulated." It's so much worse than he says. Oil companies basically own the whole Gulf region.

Republican tea-bagging superstar and knucklehead Senator Jim DeMint is putting a hold on a bill that would let the commission investigating the blowout subpoena needed information. DeMint says, "When Obama says, 'Yes we can,' we'll say, 'No you won't.'" He doesn't care "what," only "not." DeMint appears to be doing his personal best to "shrink government," and government could hardly get smaller than having senators like him. When men like him cast large shadows, it must be pretty late in the day.

❧

When a certain tugboat captain—a man I've known for several years—comes ashore after an extended stay in the waters of the open Gulf, he's got a few frustrations to relay.

"We were never given a big-picture speech about what we were all trying to accomplish," he says, "where we each fit in. It was just, like, Okay, go out to here and drive around for a while."

"Nobody really knew what they're doing," he continues. "It was very half-assed. The guy in charge of our task force, he's a nice guy, but he was lost. He's got all these shrimp boats and charter boats, and we tried to organize them into search patterns so we could know what we did. We said to the guy, 'Do you want us to give these boats courses and legs to run searches on?' And he's like, 'Sure, if you know how to do that.' He added, 'I have no idea; I'm a shore-based guy.' As far as running an effective vessel group, he was overwhelmed. I don't think he was trained. When we started talking about 'search pattern grids' and stuff like that, it was right over his head.

"We were towing four 50,000-barrel-capacity double-hulled barges. Our job was to be a receptacle for skimmed oil. The need for us was wildly optimistic. With 200,000-barrel capacity, we came back with about 1,000 barrels of oil-water mix.

"The oil we traveled fifteen hundred miles to help recover was mostly not recoverable. We saw little patches here, little blobs there, pancakes, palm-sized pieces, pennies, dimes, nickels everywhere. Most of the stuff is just broken up into tiny droplets. And it's in suspension, so there's no way you can skim it.

"So we're constantly driving around looking for something we may be able to do something with, and almost never finding it, not really accomplishing anything. I'm sure we used more oil as fuel than the amount of oil we recovered.

"On the radio you'd hear frustration from other boats. Southerners attracted to working on boats and oil rigs have a very powerful work ethic; nobody wants to feel like they're wasting their time.

"To compound it, every time we find a large enough mass to actually be able to do some productive skimming, they just hit it with dispersants. One day we were in water with a heavy sheen on top. The air stank of the crude, and there were millions of little blobs of what has become known as 'peanut butter,' or weathered crude. It was also scattered throughout the water vertically. I couldn't tell how far down it went. It lasted mile after mile, as far as the eye could see.

"This is my second major spill response—I was one of the volunteer idiots power-washing rocks after *Exxon Valdez* back in '89—and as bad as *Valdez* was, the scale of this one simply takes my breath away.

"In this horrible mess was the biggest herd of dolphin I've ever seen, about sixty to ninety, I estimated. They stayed with us for about two hours, until I changed course. As the day progressed the oil got heavier and heavier, and by sunset I thought that the next day we'd get to do some real skimming and recovery.

"About four in the morning they called and told us all to get out of there and go north of twenty-nine degrees, ten minutes latitude. They kicked all the boats out of the area by dawn, then they bombed the whole area with dispersants from planes. I said, 'What the hell do they want to do that for? We finally found it in a concentration we might have picked up, and these assholes go and spray it.'

"I don't know if those dolphins were around for that or not. I don't know if the pilots would have aborted the mission if they had spotted them in the target area. Probably not, I imagine. I only hope the dolphins somehow knew what was going to happen and got the hell out of there. I don't even want to think about what would have happened to them if they didn't."

❧

Splotches of crude are still washing up here and there along Louisiana's coast, but in early August NOAA releases a five-page report called "Deepwater Horizon/BP Oil Budget: What Happened to the Oil?"

It's got a nifty pie chart that describes the fate of the estimated 4.9 million gallons of oil, broken into seven categories. Designed to be simple and communicate clearly, it becomes a major public relations mess because: (1) some high-ranking government officials apparently can't read; (2) that's partly because there is one major thing in the report that really is pretty confusing; and (3) some reporters also apparently can't read. And what the confused officials say confuses the media totally.

Here's what the report says; you can read it for yourself:

> In summary, it is estimated that burning, skimming and direct recovery from the wellhead removed one quarter (25%) of the oil released from the wellhead. One quarter (25%) of the total oil naturally evaporated or dissolved, and just less than one quarter (24%) was dispersed (either naturally or as a result of operations) as microscopic droplets into Gulf waters. The residual amount—just over one quarter (26%)—is either on or just below the surface as light sheen and weathered tar balls, has washed ashore or been collected from the shore, or is buried in sand and sediments. Oil in the residual and dispersed categories is in the process of being degraded. The report below describes each of these categories and calculations. These estimates will continue to be refined as additional information becomes available.

The report explains what "dissolved" means with this line: "molecules from the oil separate and dissolve into the water just as sugar can be dissolved in water." Just as sugar. Isn't that nice? Facts aside, that reassuring tone makes many people feel they're being snowed. And that undermines credibility by keeping people on their guard. It's not that NOAA's scientists are giving the wrong information; it's that they're striking the wrong tone.

"I think it is fairly safe to say," remarks White House spokesman Robert Gibbs, "that many of the doomsday scenarios that we talked about and repeated a lot have not and will not come to fruition."

That's easy to say if you're not living it. Psychologically, his statement is a serious miscalculation. Most people in the Gulf want empathy, not reassurance. They want to know that their government *cares.* That's an emotion, not a desire for facts or official opinions. And because the most lingering bad taste that the nation had after Katrina was that the Bush administration seemed not to care enough, the Obama people should have understood that it was more important to show concern than to show pie charts.

Gibbs should have avoided the temptation to sound an "all's well" so early, and waited for academic experts to determine what has or hasn't come to pass *after* a relief well does its thing, when the blown-out well is officially declared good and dead. The federal government's most visible officials are misjudging people's need to *grieve.*

Even worse, some of the political folks don't seem able to read a simple pie chart. High on the list of those needing personal blowout preventers is Carol Browner, director of the White House Office of Energy and Climate Change Policy, who blurts: "The vast majority of oil is gone."

Just like that. She's totally wrong. And many people rightly jump all over her. Browner's glib comment, which rises to the definition of plain stupid, utterly undermines—yet again—public confidence in what "the government" (defined with a broad brush dipped in crude oil) is telling us.

So, for the benefit of Ms. Browner, let's review what NOAA's report actually says: "burning, skimming and direct recovery from the wellhead removed one quarter (25%) of the oil." That means, according to this estimate, that human intervention took only a quarter of the oil *out* of the system. "One quarter (25%) of the total oil naturally evaporated or dissolved . . ."

Whoa, wait a minute. Evaporated *or* dissolved? Those are very different things; evaporated means it went bye-bye in the sky; dissolved means it's asleep in the deep, still very much in the Gulf. Lumping together two very different categories that collectively account for 25 percent of the oil is another big blunder that helped make this seem very confused.

". . . and just less than one quarter (24%) was dispersed (either naturally or as a result of operations) as microscopic droplets into Gulf waters." The term "naturally dispersed" refers to oil that shattered into fine microdroplets from the sheer physical forces of being shot from a hot well into cold seawater. Twice as much is estimated to have naturally dispersed—if you call that natural—than was dispersed by chemical dispersants.

"The residual amount—just over one quarter (26%)—is either on or just below the surface as light sheen and weathered tar balls, has washed ashore or been collected from the shore, or is buried in sand and sediments. Oil in the residual and dispersed categories is in the process of being degraded." Here again, the report is confusing because oil that's been collected is no longer in the Gulf system, but at least with this category you know that the amount of residual oil in the Gulf is *not more* than they're estimating. You do know that, right?

So how much oil are they saying is in the Gulf now? Well, let's see: there's the 24 percent that's dispersed, the (up to) 26 percent that's either on or near the surface or washed ashore or is buried, plus whatever is the "dissolved" part of the 25 percent that's "evaporated or dissolved." That means the pie chart is telling us that up to 75 percent of the oil is in the Gulf. And if more of it evaporated and we split the difference, it's saying that perhaps two-thirds of the oil is still in the Gulf.

But so annoyed and upset are a lot of people—especially by the White House misstatements—that various independent scientists want a do-over. Louisiana State University oceanography professor Jim Cowan tosses the pie chart into NOAA's face, saying, "It looks like a nice neat diagram, but I have no confidence in it whatsoever." The Georgia Sea Grant program, itself part of an NOAA-sponsored university network of ocean and coastal researchers, releases an "alternative report" claiming that most of the oil that leaked into the Gulf is still present and estimating that between 70 percent and 79 percent of the oil remains in the Gulf ecosystem.

And yet, look: that's not too far off from what the original pie chart says.

Some people don't seem to care how much oil is where. Ronald J. Kendall, who directs Texas Tech University's Institute of Environmental and Human Health, says, "Even if all the oil were gone tomorrow, the effects of the spill on species such as sea turtles, bluefin tuna and sperm whales may take years to understand."

Trying to keep the message on point, the report's lead author, NOAA chief Dr. Jane Lubchenco, notes, "No one is saying that it's not a threat anymore. I think the view of most scientists is that the effects of this spill will likely linger for decades." She observes, "There's so much noise out there now saying the Gulf is dead or the Gulf will come back easily. The truth is in the middle."

Despite her attempts to recenter the discussion, the media reports on this are all over the place. One leading newspaper says, "Roughly one-third of the oil that gushed from the wellhead is out of the system: recovered directly or eliminated by burning, skimming, or chemical dispersion operations." Nope, wrong; dispersed oil is very much *in* the system.

The Associated Press, a stalwart of steady-as-she-goes reporting, publishes an article with this strikingly sarcastic (and inaccurate) headline: "Looking for the Oil? NOAA Says It's Mostly Gone."

The *New York Times* writes that federal officials are saying that only about 26 percent of the oil is still in the water or onshore. (No, see above.) Britain's *Independent* says, "Only about one-quarter of the oil remains as a residue in the environment, according to the US National Oceanic and Atmospheric Administration. The other three-quarters no longer poses a significant threat to the environment, the NOAA scientists said." The writer knows that the NOAA people said no such thing; in the same article he *quotes* Dr. Lubchenco as saying, "Dilute and out of sight does not necessarily mean benign, and we remain concerned about the long-time impacts both on the marshes and the wildlife but also beneath the surface."

Another British newspaper, the *Guardian*, says that NOAA scientist Bill Lehr "appeared to contradict the official report that he wrote" by saying, "Most of the oil is still in the environment." The paper goes on: "His statement is bound to deepen a sense of outrage in the scientific community that the White House is hiding data and spinning the science of the oil spill." But the report says what he said: that most of the oil is still in the environment. It's the media that's getting it wrong and spinning it.

Independent and academic scientists label the report misleading. "The oil has not left the building," says Ian MacDonald of Florida State University. He says the federal report gives the impression that most of the oil is no longer in the water because it was dissolved, dispersed, or degraded. "That only means it's still in the ocean, but in different forms," he points out. "It's still in the water."

University of Miami marine scientist Jerald Ault points out, "All those toxins that were injected into the Gulf, and remain in the Gulf, can be deadly to eggs and larvae and the young life stages of these species like giant bluefin tuna, yellowfin tuna, the billfishes and marine mammals, and many others. When you inject that volume of oil and dispersants into this life web, changes will echo through the system for a very long time. This is a long way from being over."

That's reasonable speculation. But it's speculation. Really, nobody knows. The stuff can be toxic, but how toxic at what concentrations, and for how long? It's continually being diluted and degraded. Its toxicity yesterday isn't the same tomorrow. And yet there must have been—I speculate—tremendous damage to sheer numbers of those eggs and larvae. We should also bear in mind, however, that the numbers of eggs and larvae are always far in excess of what the system can support. The competition and struggle for existence is so intense that under normal, healthy circumstances, only one fish egg in millions wins the lottery ticket for becoming an adult. That is where a lot of the resiliency comes from. There may be enough survivors to let the Gulf recover quickly.

So yes, this is a long way from being over. But I think it's the people and communities that will have the longest and hardest time recovering. Again, that's my speculation. But I'd bet on it.

Patches of oil are still washing up in the marshes and coastal areas of Louisiana, tourists remain skeptical and elusive, and waters remain closed to fishing.

"For technological disasters, unlike natural disasters, we see long-term impacts to communities, families, and individuals," says University of South Alabama sociologist Steven Picou. He has studied the impacts of both the *Exxon Valdez* spill and Katrina. In a natural disaster, he says, "People quit blaming God, usually after two weeks; then they come together with purpose and meaning to rebuild, so your social capital in a community grows after a natural disaster." But, he notes, "What we found in Alaska was that communities tended to lose their social capital. Their trust in local institutions and state and federal institutions, and their social networks, tends to break down. Everyone is angry and people get tired." The resiliency coast residents showed after Katrina may help them overcome the blowout, too, he says—but this will be a marathon, not a bounce-back.

And as for misjudging people's emotions, few can rival BP, which is saying it might someday go back to Plan A and use the well for commercial purposes. Tony Hayward said in June that the reservoir was believed to hold about 2.1 billion gallons of oil. Roughly 200 million gallons have leaked out, leaving about 1.9 billion gallons, over 45 million barrels. At the current per-barrel value of $82, what remains is still worth $3.7 billion. Now BP's chief operating officer, Doug Suttles, is saying—and you have to wonder why in the world he thinks it necessary to bring this up before the well is even deemed fully secured—"We're going to have to think about what to do with that at some point."

It happens to really bother some people that a company with revenues of $147 billion in the first half of 2010 would commercialize a grave site. A fifty-four-year-old real estate agent from Mississippi, her voice cracking, says, "People died out there on that rig. They can find another place. Leave that one alone."

BP still has lease access to a roughly three-by-three-mile block of seafloor there. They'll be back.

In another stroke of public relations insight, BP is now hedging about how the relief wells will be used. BP officials had insisted for months that the relief wells were the only surefire way to end the oil leak, but now they're saying that what they've already done might do the trick.

BP is now refusing to commit to pumping cement down the relief well and into the bottom of the blown-out well. But why? Maybe their reason is a good one. Or maybe they think they'll use it to produce oil. Or they want to save a little money. Thing is, we don't know. They're so inept at PR, they manage to murk up the plan that for months they've touted as their best way to finally bring closure to the blown-out well.

But Thad Allen will have none of BP's vacillating. He makes it clear that the gusher will have to be plugged from two directions to be sure it's permanently dead. "I am the national incident commander and I issue the orders," he insists. "This will not be over until we do the bottom kill." He adds, with some urgency, "The quicker we get this done, the quicker we can reduce the risk of some type of internal failure." He assures us that a relief well will inject cement into the well deep below the seafloor. "There should be no ambiguity about that," Allen says. "I'm the national incident commander and this is how this will be handled."

"I wouldn't put it as 'government versus BP,'" says one of BP's interchangeable vice presidents. "This is just about some really smart people debating about what's the best way to do things."

When people refer to themselves as really smart, my confidence in them—if I have any—declines.

Perhaps sensing certain limits, BP's CEO now (again) confirms that BP plans to use the 18,000-foot relief well to seal the blown-out well with drilling fluid *and* cement.

But stay tuned.

Maybe BP is getting distracted by the 300 lawsuits piling against it like snowdrifts. Transocean, close behind with 250 lawsuits, claims it's not responsible and has the gall to ask a court to limit its liabilities to a piddling $27 million.

Central to BP's legal strategy will be the need to rebuff claims that the company acted with "gross negligence." The difference between "gross negligence" and regular garden-variety negligence for BP, in this case, could be more than $15 billion in additional civil penalties under the Clean Water Act. Consequently, BP does what any negligent company would do: blames its partners. "Halliburton should have done more extensive testing and signaling to BP," says BP. To which Halliburton retorts, "The well owner is responsible for designing the well program and any testing."

While the principals engage in a foot-eating contest and continue to antagonize one another, the media, and us, we have a bit of luxury in asking the real questions: How much? How long? How bad? A big part of this—maybe all of it now—turns on the questions, How toxic? What will die?

Easy questions, hard to answer. People are still measuring, analyzing, writing up their findings. A fuller picture hasn't yet emerged. And so there's a tug-of-war between those who see the changing Gulf situation as a glass half empty, those who see it as half full, and those who see it as half-assed.

We can begin with brief comparisons. Oil from the *Exxon Valdez* remains obvious in the sands of Prince William Sound. Oil spilled four decades ago in a well-studied Cape Cod marsh lingers a few inches below the surface. But the Ixtoc blowout sent more than 3 million barrels into the Gulf of Mexico. That's a lot; it's more than half the total of the 2010 blowout. Ixtoc did a lot of damage, yet by most accounts, most things were pretty much back to normal a few years after Ixtoc blew.

Because an estimated couple thousand barrels of oil enter the Gulf daily from thousands of small natural seeps, the Gulf is well populated with bacteria that can eat oil. (Thank God for evolution.) Like a living inoculation, their existence gives the Gulf some powers of natural recuperation, even from such an enormous shock as this blowout. They're part of the Gulf's resilience, its flex. Microbiologist Ronald M. Atlas, formerly a president of the American Society for Microbiology, says, "I believe that most of the oil will not have a significant impact. That's been the story with spills that stay offshore." Texas A&M University professor emeritus Roger Sassen says that because the Gulf is "preadapted" to crude oil, "The image of this spill being a complete disaster is not true."

One way of seeing if invisible oil-eating microbe populations are active and growing is to measure oxygen, since microbes use up oxygen. So one research team notes that oxygen levels haven't declined much, implying that the oil down deep is degrading slowly (it's cold down there—40° Fahrenheit). Until more puzzle pieces start coming in, it's a bit puzzling.

Everyone agrees that oil is toxic, but the plain truth is, no one can say how toxic it is out in the Gulf. That's because the concentrations vary from place to place and continually change. Laboratory tests are usually done by putting organisms such as shrimp or fish larvae in a mixture of water and the chemical they're testing, and seeing what concentration kills half the organisms in forty-eight hours. Researchers want results, so the concentrations are fairly high. But they want acute results—death. Doing experiments on the effects of very low concentrations means needing more samples and much more time, which translates into money. Too expensive, so seldom done.

And real ecosystems are so much more complicated that a laboratory experiment doesn't help us understand the fate of chemicals in the Gulf's living communities. If a research team finds oil and dispersant chemicals in the Gulf at concentrations a hundred times lower than the concentrations that killed half the larvae during two days in the laboratory, does that mean half the shrimp and fish larvae will die in two hundred days instead of two? No, it doesn't. But that also doesn't tell you whether the lower chemical concentrations will hamper the organisms' ability to continue finding food, grow normally, avoid predators, migrate, fight off infections, and so on. So when a scientist says, "We don't know how toxic it is," that's the truth.

But let's ask it another way. Let's look at trends. Is stuff dying? What isn't dying? What about recovery—are any components of the Gulf or its coastal marshes showing signs of bouncing back? Or not?

Now, here's the thing: though there's still a lot of disagreement, suggestions that the oil is rapidly disappearing are beginning to come from various independent sources. Even before July rolls over to August, the floating oil mats that covered thousands of square miles of the Gulf are largely gone. Remaining oil patches are quickly breaking down in the warm surface waters.

"Less oil on the surface does not mean that there isn't oil beneath the surface, or that our beaches and marshes are not still at risk," says NOAA chief Dr. Jane Lubchenco. "The sheer volume of oil that's out there has to mean there will be some very significant impacts. Dilute does not mean benign."

That's true. But it might also be fair to bear in mind that dilution is what pollution passes on the way to benign. In other words, the trend seems good.

My friend the tugboat captain is heading back to shore from another trip into the Gulf. This time he's off Florida, approaching Pensacola. And this time he's got good news: he's seeing numerous schools of the tuna cousins called little tunny. Because he's driving a boat in open water, he's got plenty of time to talk.

What are his impressions?

Despite the fish, "a lot of the water just looks off," he says. "What it usually looks like and what it looks like now are two different things. I know that's very subjective. I haven't seen any obvious slicks recently; it just looks off. It's like an old piece of Plexiglas; it just doesn't have that clarity like when it was new. Ever so slightly opaque.

"And I've pulled up a bucket of water that seemed perfectly clean, but it feels slimy. Like I say, the closer to the well, the worse that was. I'm accustomed to seeing large areas of sargassum. I've seen none of that. What you see is little bits, like it was run through a food processor, nowhere near the amount I would normally see, and no big patches. And I'm looking for it on purpose. The coloring is usually bright yellow, but it was gray; that's something that really jumped out at me. Sea turtles, a few, but nowhere near what I'd expect.

"And once or twice a day, twenty to sixty miles out, I saw isolated large shrimp swimming on the surface. Not near a weed line or near any protection. And also blue crabs, way offshore on the surface. I've never seen shrimp on the surface, and I've never seen blue crabs so far offshore. These were not small shrimp; I could see them from the wheelhouse, clearly alive, snapping their tails for locomotion. No other shrimp with them and no shelter, so what they were doing there, I've got no idea. I've seen this about a dozen times in the last ten days. I saw no sign of obvious distress. They were just very much out of place. What to attribute it to, I don't know. Except we've had this enormous event, and then you see behavior you've never seen before.

"And I could count on one hand how many flyingfish I've seen. That's extraordinary; I'm used to seeing them everywhere. They've got to be one of the more vulnerable animals to the oil.

"The dolphins, they're there, but I have seen, I'd guess, less than a third of what I'd expect, and other than that one enormous herd I saw the day before they dropped dispersants near us, they're pretty scarce and the groups are smaller. They seem more common to the east, away from the leak site. I haven't seen a lot of seabird activity except farther east near Pensacola, where the little tunny are. And I only saw this once, but the action was fast and furious: two or three spinner sharks leaping out after prey—spectacular—with multiple twists before they splashed down."

When I ask if he's going back out, he tells me he's not sure. "I don't know if they're going to keep us on the job much longer," he says. "We're not doing anything. It's just wasteful to keep us on the payroll. They should just give the fishermen the money instead of paying us to run in circles."

There's a pause on the line—it's not like him to stay quiet long—and then he adds this coda: "I can honestly say I have spent the whole summer doing absolutely nothing productive at all, and burning an enormous amount of fuel in the process. We just put a lot of miles and a lot of hours on the boat, for absolutely nothing. We accomplished no work, helped no one, produced no seafood, saved no lives, nothing. The only thing I have from all this is, at least I got to see it."

On August 10, NOAA reopens federal waters off a portion of the Florida Panhandle. The modification applies to fish only. No crabs. No shrimp.

Fraud alert: BP discovers that the numbers of commercial fishing licenses sold since April is up by 60 percent compared with last year. Fishing was *closed.* "There was an approach by two individuals asking me to sign that they had worked for me, that they had been deckhands for me," says one boat captain. "I had never seen these two individuals before in my life." One guy charged with filing false public records and theft by fraud faces possible prison time, large fines, and even hard labor if convicted. But like dead birds, for every one you find, there could be ten out there.

Legal alert: BP is prepping for its rootin' tootin' rodeo of legal wrangling. University of South Alabama's Bob Shipp says BP's lawyers tried to hire his whole Department of Marine Sciences to do research for them. "They wanted the oversight authority to keep us from publishing things if, for whatever reason, they didn't want them to be published," Shipp says. "People were to be muzzled as part of the contract. It's not something we could live with."

Tulane law professor Mark Davis reminds us that BP is doing what other oil, tobacco, and pharmaceutical companies have done in the past: hiring scientists to do research they want kept secret. "When the best scientists have evidence that would work against you, but they're not able to present it to the public, well then, you've essentially bought some silence."

LATE AUGUST

Relief well? Cement? No cement? Just a week ago BP stopped hedging on whether it would send cement into the bottom of the blown well. Its people have reconfirmed that yes, they will cement the thing using the relief well they've been digging for three and a half months. Our adamant commandant the Thadmiral has resolutely insisted that he gives the orders and that the endgame equation is: relief well = bottom kill = pumped cement = final victory.

Now, what's this? BP *and* the feds are back to wondering whether the bottom kill is even necessary.

The Thadmiral can't really be telling us that the blown-out well may not need to have heavy fluid and cement pumped into it through the relief well after all, can he? I guess he is. He's saying, "A bottom kill finishes this well." Then he adds, "The question is whether it's already been done with the static kill."

I don't get it.

This statement seems—what's a good word?—it seems *evasive.* How could no one have even hinted at this possibility all along? What's *up?* The relief well is late, they've been hemming and hawing, and now this.

It appears that after the relief well intersects its target—assuming it does, eventually—BP wants to check to see whether the cement pumped in through the top of the blown-out well went all the way down the casing, went out through the bottom of the well, and came up far enough between the casing and the rock to have plugged the whole shebang to secure satisfaction.

If so—spokesmen now say—sending cement in near the bottom might not be necessary. But as the Associated Press puts it, this new idea "would be a hard sell to a public that's heard for weeks that the bottom kill is the only way to ensure the well is no longer a danger to the region."

Exactly.

What is Allen thinking? The Thadmiral has zero to gain by even introducing this as a question now. It once again undermines the perceived credibility of everything he and other federal officials have been saying for months—even if what he's saying is perfectly reasonable. Sometimes you just gotta do what you've been promising.

I admire Thad Allen's willingness to reassess and to change course accordingly. It's a sign of intelligence. A foolish consistency indicates a small mind. He may be right, and he's certainly being reasonable.

But people don't want to feel reasoned with. They want to feel safe. This whole disaster is psychological almost as much as it is physical—and America hates flip-floppers. To suddenly open this as a question scratches the scab just forming over this gaping wound. For all of us who've been assured for months that a relief well that pumps cement is the only assurance, it feels excruciating.

Emotions. The public mood is at least as big an issue as the oil. I don't think the officials have realized this. If they have, they haven't dealt with it skillfully.

Emotions dominated the region while the oil flowed. Yes, there is real blame for what caused the blowout and for the utter unpreparedness to anticipate and deal with a blowout one mile deep. Some places, like Grand Isle, received awful coatings of oil. Wildlife suffered, and there is damage to natural habitats (though, luckily, less than feared).

But the two most devastating social and economical consequences of the blowout—the region-wide tourism meltdown, the vast fisheries closures—have been, in much of the wider region, emotional responses to aesthetics and perceptions, rather than necessary responses to real dangers. That's been true even where little oil reached. Panic, blind anger, and the months-long inability to know how long the oil would continue flowing have dominated people's responses to the event—including my own—especially in the months from late April until early August, when the leak was finally stopped.

Now that the oil has stopped flowing, we can begin to assess the extent of the event, the damage done, the prospects for recovery—and we can begin to calm down.

We've seen what caused this well to blow out, and the varied responses during the chaotic months while the oil was flowing. Now we can afford ourselves a little more calmness, clarity, rationality, and insight.

So what is the main observable effect to date, and what has taken the biggest hit—marshes, fishes, birds, water quality? It doesn't seem that way. Many Gulf Coast beaches are now free of both oil *and* tourists. The tourism industry projects losses in the $20 billion range. Both the tourism and the seafood industries see themselves as battling negative perceptions more than actual oil at this point, and perception may turn out to be the most costly of consequences.

There's still a lot of oil out in the Gulf. The twin fears for deep-sea oil plumes have been that billows of toxic hydrocarbons would roll through plankton communities, causing massive damage, and that the dissolved oil and gas would trigger a population explosion of oil- and gas-eating microbes that would burn up most of the surrounding oxygen, strangling nearby life.

NOAA and the EPA now report on dissolved oxygen at some 350 sampling stations. Bottom line: no oil-caused dead zones developed; none are expected. The average normal oxygen level in the Gulf at plume depths is 4.8 milliliters per liter. The average in-plume level was 3.8. The depressed oxygen level indicated bacteria eating oil.

But for the water to be a dead zone, the level would have to be below 1.4. Modeling indicates that dead zones have not developed because of oil mixing with nearby water. In other words, it's a relatively small amount of oil in a big ocean.

"We're seeing very few, if any, moderately or heavily oiled turtles," says sea turtle biologist Blair Witherington. Until the blowout was capped, the floating mats of sargassum weed that young turtles must spend years in had collected their own mats—of oil. Many young turtles must have perished unaccounted for. But the sargassum habitat is revitalizing, the oil dissipating from it. The blackened sargassum of early summer is being replaced by clean new growth teeming with crabs and other recovering creatures, and accompanied by turtles either clean or so lightly oiled that they can be cleaned on the spot and released. Says veterinarian Dr. Brian Stacy, "I personally didn't anticipate such a dramatic change so quickly."

Louisiana State University professor emeritus Edward Overton observes, "The Gulf is incredible in its resiliency and ability to clean itself up. I think we are going to be flabbergasted by the little amount of damage that has been caused by this spill."

But remember the oil and dispersant droplets in the larval crabs, showing that the spill was already climbing up the food chain? A pretty explosive find. Yet we haven't heard more about it. So a reporter from the *Orlando Sentinel* followed up, asking one of the scientists about it.

"That was a mistake," she said.

But what about the droplets seen in the crab larvae?

"We don't know what it is," she said. "It could be natural."

The owner of a bait-and-fuel dock peers into the marina's clean-looking water and says, "The spill isn't as bad as the media has suggested."

A charter boat captain who's been fishing several times a week since his area reopened says, "The perception is, everything down here is absolutely slap covered in oil. But that's not true. You could drive around all day and not find it." He's back to catching redfish and speckled sea trout.

Even in Louisiana's Barataria Bay, where two months ago gruesome scenes of oil-clotted pelicans horrified America, by mid-August green shoots began appearing in oil-blackened grass and mangroves and cane started to regrow. The Gulf region does indeed appear to have escaped the most dire predictions of spring. It could have been a lot worse.

Of the oil that reached the coast, most was stopped by the marsh itself. Marshland closest to the Gulf took the worst of the spill, absorbing oil more thoroughly than all the boom in BP's checkbook, and preventing it from moving farther into thousands of square miles of marshes. Estimates of how much of Louisiana's vast marshes have been affected are remarkably small. The Associated Press calculates that, over hundreds of miles of Louisiana coastline, at most 3.4 square miles of marshland got oiled. There are roughly 7,000 square miles of marsh.

It's tempting to shout, "The marsh is coming back!" But it's just that the vegetation is regrowing. The *marsh* is disappearing.

In Shell Beach, Louisiana, a man saw me staring at all the dead trees standing in the marsh. "When I was a kid," he said, "this channel wasn't here. And all of this was woods, all cypress and oak and other hardwoods. Everything behind us was cattle pasture. Now it's water. The channel let salt water pour in. Now you see all these dead trees. Those all died in the last ten years. The road leading up to the bridge you came over? That road was canopied with trees. It was real nice. So it gives me bad memories. Everything tries to live. Two feet above the water, a tree will grow. Everything wants a chance to survive. The land is subsiding. Anyway, yeah, it was a paradise. Everyone got told they were all gonna be rich when the channel came. Look at any other state. Their coasts continue to thrive. This one died."

"We were anchored," writes Louisiana outdoor columnist Bob Marshall three weeks after the leak has been stanched, "watching redfish push wakes in clear water as they raced along a bank lined with very green and very healthy *Spartina* marsh. Shrimp were leaping from the water in attempts not to become redfish dinners. Blue claw crabs were riding the outgoing tide toward the Gulf of Mexico. Pelicans were diving on mullet schools. Mottled ducks were puddle jumping, and sand flies were taking their pint of blood from my ankles.

"One of the planet's most vibrant and dynamic ecosystems seemed the picture of health. Of course, like most wetlands sportsmen, I knew better. That's because I know these marshes are turning into open water at the rate of 25 square miles per year."

The delta originally covered something like 8,500 to 10,000 square miles, and has lost about 20 percent of that area, or very roughly 1,800 square miles of marsh, with recent loss rates around 20 to 40 square miles a year. All these estimates vary, as do the actual loss rates. Hurricanes Katrina and Rita alone dissolved over 200 square miles of marsh to open water, but much of that had been already degraded marsh. The Mississippi River is the main source of the delta's nourishment, but engineering projects have almost completely isolated the river from its own delta. The projects include levees built for ships and against floods, thousands of miles of channels sliced through the mazes of marshes, and the dredging and deepening done for shipping, and oil drilling, and the ships and traffic buzzing to and fro to service the rigs. The delta is disintegrating because the sediment that washes from the heartland now gets shot straight out into the open Gulf. It never gets a chance to build the marshes. Oil and gas pumping have also helped the marshes subside. The rise in sea level isn't helping.

A lot of marsh remains, but the amount lost—and with it the lost wildlife, recreation, and fisheries productivity—is staggering. The marshes are one big reason the Gulf produces more seafood than any other region in the lower forty-eight states. But some estimates say they'll largely vanish by 2050.

A certain irony is not lost on Dr. Felicia Coleman, who runs Florida State University's Coastal and Marine Laboratory. "There's a tremendous amount of outrage with the oil spill, and rightfully so," she notes pointedly. "But where's the outrage at the thousands and millions of little cuts we've made on a daily basis?"

BP has indeed been the subject of national rage all summer. One of the greatest fears was that the oil would destroy a great swath of marsh. Yet the things that are really destroying the marsh are all intentional. They could be fixed.

BP may end up saving more wetland than the oil ever harmed. The new national focus on the Gulf has helped bring attention to the delta's disintegrating marshes. Some of the billions BP is expected to pay in fines could bankroll restoring critical wetlands and reengineering projects that Congress hasn't bothered to fund. The Obama administration seems to grasp this. Navy secretary and former Mississippi governor Ray Mabus, having been tasked by the president to draft a long-term restoration plan for the Gulf, says he envisions spending some of the money from BP's anticipated fines on repairing wetlands. In one sense, it's a way for BP to give back for using all those wetland-killing canals and ship channels dug throughout the marshes to serve the oil rigs.

On August 16, Louisiana shrimpers go back to work—their real work, not dragging useless booms. Seventy-eight percent of federal waters—up from 63 percent—are open. On August 18, wildlife rehabbers release the first turtles back into the Gulf. And on August 20, the Louisiana Wildlife and Fisheries Commission reopens recreational fishing in all state waters that had been closed.

If that doesn't help people's mental health a little bit, BP announces that it's going to give $52 million to five behavioral health-support and outreach programs. "We appreciate that there is a great deal of stress and anxiety across the region, and as part of our determination to make things right for the people of the region, we are providing this assistance now to help make sure individuals who need help know where to turn," says a BP spokesman.

BP's $52 million for mental-health care doesn't by itself solve the material basis of problems like sleeplessness, anxiety, depression, anger, substance abuse, and domestic violence. To put it plainly, money will be the only anxiety cure for people like Margaret Carruth, who lost her hairstyling business and then her house when tourists stopped coming and locals cut back on luxuries like haircuts. Now she sleeps on friends' couches or on the front seat of her blue pickup. For people like her, the blowout continues to wreak its damage full-force.

In fact, people's mental health may get worse as the financial strains persist. A Louisiana husband and wife who both work in the seafood business and have seven children say they've received only $5,000 in claims payments since May. One single mother of four who worked as a sorter on a shrimp boat used to earn about $4,000 a month. Her BP payments: less than $1,700 monthly. "I worry about my kids seeing me this way," she says, "and them getting sad or it affecting their schoolwork."

At a day-care center in Grand Bay, Alabama, preschoolers whose parents were left jobless by the blowout are lashing out, their caregiver reports, by "throwing desks, kicking chairs. It's sad. With this," she continues, "people do not have hope. They cannot see a better time."

Top kill. Bottom kill. Junk shot. Dome. Riser. Skimmer. Annular. Cap. Relief well. Negative test. Blind shear ram. Centralizer. Drilling mud. Spacer. Blowout preventer.

None of those words refer to a woman sleeping in her truck or a child throwing a desk. And yet, of course they do.

For the rest of us, here's a new reason to seek mental help: the Thadmiral says he's no longer giving timetables for the final plugging of the now-quiescent well. He says he'll give the order to complete the relief well operation when he is ready. He doesn't want to give timelines anymore because, he says, if he has to change them—*it could cause a credibility problem.*

On August 27 NOAA reopens to commercial and recreational fishing over 4,000 square miles off of western Louisiana. Ten times more than that remains closed, about 20 percent of the federal waters in the Gulf. At the disaster's height, 37 percent of Gulf federal waters—88,000 square miles—were closed.

Meanwhile, with much fanfare, Canadian prime minister Stephen Harper announces a new protected area for beluga whales at the mouth of the Mackenzie River, in the Beaufort Sea. The Beaufort Sea region is home to one of the world's largest summer populations of belugas, which go there to feed, socialize, and raise their calves. In the remote town of Tuktoyaktuk, Northwest Territories, Harper proclaims, "Today we are ensuring these Arctic treasures are preserved for generations to come." But the government has already granted licenses to two companies that have proven the existence of about $6.6 billion in oil and gas reserves next to the beluga protection area. And the government will allow the companies to "continue to exercise their rights." Member of Parliament Nathan Cullen says with apparent disgust, "This is a protected area that protects nothing except oil and gas interests. It's some insane notion that we can draw a line in the water and drill right beside it."

Taking a leaf—as the saying goes—from Obama's book, Indonesia demands $2.4 billion in compensation from a Thai oil company for a blowout in the Timor Sea that lasted three months, spread a 35,000-square-mile slick, and was eventually plugged with a relief well. Indonesia says the oil damaged seaweed and pearl farms and hurt the livelihoods of 18,000 poor fishermen.

Taking a leaf from Exxon's book, the Thai-owned oil company rejects Indonesia's demand.

There's been a lot about BP to quite rightly criticize. All the good money it's now throwing around is certainly self-serving. We are right not to let the corporation simply buy our goodwill; in important ways, it doesn't deserve it. But when, in the wake of the *Exxon Valdez* spill, the Congress of the United States of Corporate America passed the Oil Pollution Act of 1990, it sent squeals of delight through the petroleum industry by capping an oil company's liability at a measly $75 million. Despite its ability to do unlimited damage, Big Oil has, ever since, operated with that congressional promise of impunity.

And so we must acknowledge that BP has voluntarily waived that protection. Compared to the way Exxon turned on the people of Prince William Sound, there's no comparison. In the wake of this blowout—and I say this with no disrespect to the families of those who died on the Deepwater Horizon—BP has done some honorable things for the people of the Gulf and America. More honorable, one might observe, than the U.S. Supreme Court, when the Court basically let Exxon off the hook. How sad is that?

EARLY SEPTEMBER

Morgan City, Louisiana. "Yes!" the sign says. "We Are Having Our 75th Annual Shrimp and Petroleum Festival."

"We still need both," says Lee Darce, assistant director of the festival. "That's what makes our community. That's our lifeblood." Mayor Tim Matte is aware that the festival can seem pretty weird to outsiders. "But we've always thought it's unusual that they think it's unusual."

Fire engulfs an oil production platform one hundred miles off the Louisiana coast. Though the fire started in living quarters and there is no loss of life, to a region whose nerves are so frayed the news comes like fingernails on a chalkboard. The same day, from Panaji, India: "Thick and dark layers of oil being deposited with each lapping wave along the sun-kissed beaches of Goa could be another ill omen for the Goa tourism industry."

On September 2: NOAA reopens fishing and shrimping in over 5,000 square miles of water stretching from Louisiana to Florida. NOAA throws open another 3,000 square miles of fishing area on September 3.

Festivals notwithstanding, there is a big difference between what fishers need and what oil companies need. Oil companies need oil. If it was on land, they'd get it there. And they did. That's why they're going into deeper and deeper water. They don't need marshes, clean water, or vibrant wildlife populations. If the whole ocean was suddenly empty of fish or the marshes vanished or the Gulf's water all evaporated, they'd still be there for the oil.

Fishermen really need the place, and they need it to be working and pretty intact.

Of about 8,500 water samples so far taken by NOAA from Mississippi to Florida, only two—both from Florida's Pensacola Pass—came back positive for oil. NOAA scientist Gary Petrae reports, "We are not finding anything, and even when we're suspicious of oil being present, we're finding that we're wrong. We're doing the best we can—and we can't find it." NOAA scientist Janet Baran says, "We haven't seen any oiled sediments." Her crews have looked at more than 100 sediment samples from federal waters more than three miles offshore, including near the well. "All the sediments we have taken," she adds, "have no visible oil on them."

That's the good news.

But while she says, "We haven't seen any oiled sediments," the University of Georgia's Samantha Joye says, "I've never seen anything like this." Joye describes layers of oily sediment two inches thick. Below the oily layer, she says, she finds recently dead shrimp, worms, and other invertebrates. And University of South Florida researchers see what they believe are droplets of oil on the seafloor. "It wasn't like a drape, like a blanket of oil, don't get me wrong," says David Hollander. "It looked like a constellation of stars that were at the scale of microdroplets."

What are we to make of such different findings? Different samples, different times, in different parts of the Gulf where heavy oil did or didn't go. Pieces of a puzzle.

Greenpeace says it's still easy to find oil just a few inches down in the sand on beaches that had been obviously oiled a few weeks back. That's believeable, and the videos look convincing.

Less convincing is that Louisiana's Governor Bobby Jindal is still trying to get his one hundred miles of sand berms built. BP has agreed to pay a hefty $360 million for them. But the Environmental Protection Agency is urging the Army Corps of Engineers to turn down the state's sand berm project, saying berms don't do anything and can harm wildlife. Ostensibly they're to stop oil from contaminating shores and marshlands. Using a May permit, the state spent tens of millions of dollars to build four miles of berms.

Here's a good line of BS: one of the governor's aides says the berms they built received "some of the heaviest oiling on Louisiana's coast." So what are they, oil *magnets?* He reportedly says the Louisiana National Guard has picked up at least a thousand pounds of oily debris from the berms. You know how much oil and sand it takes to make a thousand-pound pile? Very, very little. Let's put it this way: one cubic yard of sand weighs 2,700 pounds. He says, "Now is not the time to stop protective measures that have proven their effectiveness." Actually, now might be a good time.

I suspect that this desire for berms stems from a fear of hurricanes, not oil. Is my suspicion misplaced? Says Grand Isle's mayor, "What is wrong with us dredging and building these islands back up?"

On September 8, BP's just-released internal investigation spreads the blame widely, declares the disaster—lawyers and jurors, please take note—a "shared responsibility," blames "no single factor," and adds, "Rather, a complex and interlinked series of mechanical failures, human judgments, engineering design, operational implementation, and team interfaces came together to allow the initiation and escalation of the accident. Multiple companies, work teams, and circumstances were involved."

Oh, and also, even after all the stuff leading to the blowout went wrong, the blowout preventer should have activated automatically, sealing the well. BP says the device "failed to operate, probably because critical components were not working."

Self-serving as all that is, it's also probably all correct. But there's "shared responsibility" and then there's "shared responsibility." It certainly does seem that multiple workers from multiple companies made multiple errors. But whether BP shares or owns "responsibility" will largely come down to legal definitions of said term. I once got a ticket for an undersized fish that was caught on my boat but that I did not catch and did not measure and did not put in the cooler myself (it was a quarter-inch short and the mismeasurement was a friend's honest mistake). But as captain, I was responsible. I got the ticket. Sometimes that's the way it is; the crew messes up, but the captain pays.

The sign at Pensacola Beach proclaims "World's Whitest Beach." And still, it pretty much is.

People here say they dodged a bullet. Workers are keeping the main beaches almost clean. Fort Pickens National Park maintains a plentiful complement of pelicans, gulls, nesting terns, herons, plovers, dolphins. So lovely. A few well-weathered tar balls here and there are easily coped with. One heron has a smudge of oil on its throat. Other than that, it looks nice and seems certain to survive.

And that worries me. On my first trip to the Gulf after the explosion, I feared the worst case would be that the blowout would ruin the Gulf's marshes and beaches and fisheries and wildlife for years to come. Now a new worst-case scenario arises: What if it doesn't? What if, having looked catastrophe in the pupils, we decide the worst blowout ever is simply not so bad? What if we think that wherever, whenever the next one comes, we can just deal with it? What if we do, in fact, hit the snooze button? And nothing changes. Then this was all for nothing; it was just an unredeemed sacrifice, an unmitigated disaster.

The thing about having dodged a bullet is: if you just keep staring, you get shot.

Engineers are preparing to start the delicate remote-control work of detaching the temporary cap that first stopped the gushing oil, so they can raise the failed blowout preventer the mile from seafloor to surface. There's concern that in the process, the crane may accidentally drop the 50-foot, 300-ton device onto the wellhead. So my question: What's the rush? Why not wait for the relief well to securely seal the well bottom? Why the hurry to do this now? Didn't we learn that hurrying wasn't the right thing to do with this well?

Allen is holding his cards tight, but the relief well must be close, right?

After four months and another recent delay for bad weather, the relief well drill bit is now churning nearly 18,000 feet beneath the rig. The task is a bit like hitting a dartboard three miles away. Except the dartboard is under a mile of water and two and a half miles of rock, and it's not a straight shot.

Drilling the final stretch is a slow, exacting process. The drillers dig about twenty-five feet at a time, then run electric current through the relief well. The current creates a magnetic field in the pipe of the blown-out well, allowing engineers to calculate distances and make fine adjustments.

To guide the relief well to its target, BP has picked John Wright, who, after four decades of work drilling forty relief wells around the world, can say, "We've never missed yet. I've got high confidence."

Well, okay; that's the kind of guy we want.

THE NEW LIGHT OF AUTUMN

On September 17, 2010, the long, long-awaited relief well—one of them, anyway—reaches and breaches the quiescent blown-out well.

"Five agonizing months," as the Associated Press puts it. The next step: drive the long-awaited cement stake deep into its black heart, plugging it up for good.

BP's website has this announcement:

> Release date: 17 September 2010 HOUSTON—Relief well drilling from the Development Driller III (DD3) re-started at 7:15 A.M. on Wednesday, and operations completed drilling the final 45 feet of hole. This drilling activity culminated with the intercept of the MC252 annulus and subsequent confirmation at 4:30 P.M. CDT Thursday. Total measured depth on the DD3 for the annulus intercept point was 17,977 feet. Operations conducted bottoms up circulation, which returned the contents of the well's annulus to the rig for evaluation. Testing of the drilling mud recovered from the well indicated that no hydrocarbons or cement were present at the intersect point. Therefore, no annulus kill is necessary, and the annulus cementing will proceed as planned. It is expected that the MC252 well will be completely sealed on Saturday. Once cementing operations are complete, the DD3 will begin standard plugging and abandonment procedures for the relief well.

Saturday, September 18. While the final cementing is under way out in the Gulf, Coast Guard admiral Thad Allen, the man the *Washington Post* called "perhaps the least-excitable person in American public life," walks into a Washington, DC, coffee shop looking so casual and relaxed that—though I've seen him on television enough times and I'm expecting him—at first I don't recognize him. He's just been mandatorily retired from the Coast Guard at age sixty, so he's out of uniform—wearing a casual short-sleeved shirt—and so he's free to accept my offer to buy him a cup of coffee and his almond croissant. Dr. Jane Lubchenco, NOAA administrator, is here too. But when I make the same offer she pats my shoulder and says, "That's nice of you, Carl, but I'll pay for my own, thanks."

Lubchenco leads us out the back of the shop, into a nice garden area with tables under the shade of a grape arbor. It's a perfect September morning. "We got lucky with the weather," she says. "Grape arbors wouldn't work so well in rain."

"First off," I say, "how much time do we have?"

"About an hour," Lubchenco answers. "I'm leaving this afternoon for meetings in Europe."

"An hour's good," Allen says.

Because the long-heralded final cement is getting pumped down the relief well as we take our seats, I'm sure Allen's got plenty of other things on his mind today. So I appreciate that on such a momentous morning, he's decided to peel off some valuable time for breakfast.

Recently, I happened to catch a long interview with Allen on *Charlie Rose.* Allen, who's beefy, with a military-style crew cut, took time to describe how much his participation in releasing several rehabilitated sea turtles had meant to him. I resonated. He seemed centered and insightful, and kind. And I started feeling an uncomfortable twinge—more than a twinge—that my summer-long simmering mental caricature of him was off base. Yes, I didn't like some things he'd said and done. But he'd worked to lead the region through two major disasters, Katrina and this. I could not think of anyone else who'd quite done that. And I was surprised to find myself thinking that if a hero is someone who steers events during a national crisis, Allen's as much a national hero as anyone I could think of. Well, *that* was certainly a startling thought. I've been critical all summer. But for everything there is a season. A time to cast stones, and a time to gather stones together.

A few days ago, Dr. Lubchenco was the surprise guest speaker at a National Geographic event honoring the famed ocean explorer Dr. Sylvia Earle. Someone tapped my shoulder and said to meet him at a nearby restaurant, and when I did, I was surprised to find myself sitting next to both Jane Lubchenco and Sylvia Earle. I knew that Lubchenco had worked a lot with Thad Allen recently, and after dinner, as we were saying our good-byes, my twinge prompted me to tell her that I'd been writing critically of him, and I was getting the feeling that this might be unfair. "Oh," she said, "that *would* be terribly unfair. He's a good guy. You should meet him."

And so as we sit down, Lubchenco, a coastal ecologist by profession, explains by way of introduction that Allen's understanding goes beyond the law he must enforce and includes the ecology, the science, the regional culture, and oil drilling technology.

"I stay up late at night studying," says Allen quite matter-of-factly. "I had to do the same during Katrina. You have to convert it to something the public can understand. If you don't put a public face on these disasters, you'll fail immediately. That's what happened during Katrina. Mike Brown went and hid, y'know—"

Katrina was then, this is now. Because we're pressed for time, I want to get down to it. So, I say, if the problem during Katrina was that the federal face was hiding, the problem this time was that the Coast Guard seemed to be guarding *against* public understanding and access. "How in the world," I demand, "was declaring a felony for getting near booms—"

"I can explain." Allen takes a sip of his coffee, a bite of his croissant, and says, "My policy was open access, except where safety and security were issues. Early on, right over the site, we had eight midair near collisions. People thought our flight restrictions were to stop the press; but it was simply to prevent collisions. Regarding boom, there were people vandalizing it to get boats in and out. That wasn't illegal. So, like you can't give a parking ticket unless you make parking illegal, we did it to have something to enforce. It was to keep people from damaging the boom. Things happen in the fog of war. But would we go and arrest someone just sitting near a boom? No. That was misunderstood."

I say, "It had a bad effect on people's perceptions—including mine—about whose side you were on."

"Yeah, I understand that," he says. "But I can tell you there was no untoward intent."

And why such heavy reliance on booms anyway? I ask. They obviously weren't up to the job.

"Oil was getting under and through the boom," Allen acknowledges. "Boom is easily defeated."

"So why—?"

Lubchenco begins, "The Oil Pollution Act directs NOAA to do research on cleanup, response, these kinds of things, and—"

"Budget casualties," says Allen, beating her to the punch line. "We stopped doing research and development, so we never advanced beyond skimming and burning and dispersants."

Lubchenco offers, "Every time they tried to put it in the budget; I wasn't there then but I'm told—"

"I was budget director for the Coast Guard," Allen asserts. "This stuff was defunded in the nineties and never got back into the budget. You can quote me on that."

I know we're pressed for time and I don't want to keep you, I say, but that gets to the other peeve: such heavy reliance on dispersant chemicals.

"This is really important," Allen explains. "Because the Oil Pollution Act of 1990 was Congress's response to *Exxon Valdez*, legislators were producing regulations aimed at tanker spills. But drilling technology was going to deep water. We lost it on the technology."

I comment that what people saw was the Coast Guard building a fire truck while the house was in flames.

"If you're faced with building the fire truck," Allen says, accepting the premise, "you can't deploy a response system that doesn't exist. You can't spend time investigating what happened. You can't get bogged down measuring what's happening if your urgent goal is to stop it. You optimize what you've got. After you're done, you analyze. We did get better over the weeks—. I don't think the response was ever optimized. The availability of tools, the weather, how fast we could get ramped up—"

Allen explains that the Coast Guard's role is to enforce and implement laws and regulations. The Oil Pollution Act of 1990 laid out a spill response called the National Oil Spill Contingency Plan. That creates limitations, but he says it's worked well—until now.

"This thing went off the scale," he adds. Allen explains that the oil spill contingency plan had protocols allowing dispersants and surface burning, and when this blowout happened, events triggered the protocols. "But," he says, "two things went off the scale: one was the total amount of oil; the other was that when the protocols were written, no one envisioned injecting dispersants at depths of five thousand feet.

"And we had what I call 'the social and political nullification' of the National Contingency Plan," he says. In other words, the public and politicians weren't buying the constraints of the law. Main case in point: "The National Contingency Plan says the spiller is the 'responsible party.' That means they have to be there with you. But having BP with us created cognitive dissonance with the public. People didn't understand; how could BP be part of the command structure? But that was what the law required."

He thinks it would be best to have a third party—not the oil company, not government—in charge of spill response. "Too much perception of conflicts of interest otherwise," he says.

Another problem: "BP's efforts to just keep writing checks to state and local governments—you can describe it any way you want—allowed those folks to act outside of federal coordination. Perfect example: the people of Plaquemines Parish decided they wanted to put boom out in Barataria Bay. That should have been coordinated with us. But they independently went out and did it. And to keep the boom in place, they hired a contractor who drove PVC pipe into marshland—every twenty feet. They did that with BP money. Obviously not good for the marsh. The law says it has to be federally coordinated, but that was hard when the law was politically, socially, and economically nullified. There were other plans to close off estuaries with rock jetties so oil wouldn't get in. But that would basically destroy the ecosystem."

"We were very insistent on preventing people from trying to clean marshes," Lubchenco adds. "That just pushes the oil into the sediment, and then later it keeps coming up." After the spill from the *Amoco Cadiz*, some oiled marshes were bulldozed; those are the areas that haven't recovered. Marsh recovery seems to depend a lot on not disturbing the plants' roots. With the *Exxon Valdez*, the mistake was to pressure-wash tidal zones with hot water. Hard clams are still diminished in washed areas because the structure of sediments was disrupted.

Following up on BP's unusual eagerness to write checks—thinking again of Exxon—I ask for their thoughts on why BP agreed to fund a $20 billion account when no law required that.

"Early on," Allen replies, "BP was saying yes to just about anything. It's safe to say BP thought the situation was an existential threat to their corporation and their industry."

"But—early on, too," Lubchenco says turning to Allen, "BP wouldn't let us get the video. That was a case where they had to be ordered to—"

"But even there," Allen replies, "I don't think they were trying to hide anything. Their position on live video was that with several remotely operated vehicles being very delicately maneuvered near each other, they didn't want the operators under the added stress of being watched real-time."

"That's reasonable," I say.

"It is reasonable," Allen affirms. "But you know what? I said no. I said, 'At this point—you lose that.' I told them, 'You've had a market failure and you're responsible. So you lose discretion and things get dictated. That's the way it is.' "

Allen adds that in dealing extensively with BP's senior management, he sensed "*no* indication of any ulterior motives other than to reestablish their credibility, do what they had to do, and be responsible."

Seeing my face turn skeptical, he continues: "Put it this way: there's a difference between what seems like foot-dragging and obfuscation—and competence. BP is an oil company; they're very good at getting oil out of the ground, but they absolutely suck at retail. One-on-one interactions with people—they don't have a clue. You cannot hire consultants to give you compassion and empathy. A lot of things people got upset about and viewed as bad intent on BP's part were, in my view, a lack of competency, capability, and capacity. But it produces inaction and people get the same perception."

Speaking of inaction and perception, I ask how Coast Guard rear admiral Mary Landry could have said—after the whole rig sank, with eleven lives—that it was too early to call this a catastrophe and that no oil was leaking. Why didn't they act immediately as if this obvious catastrophe would produce a worst-case blowout, and hope they were wrong?

Lubchenco insists that this is what they did. That the president said in his very first briefing, "I want everybody prepared for worst-case scenarios"; that within hours, her agency began generating models of currents, predicting where oil would be going. "Our folks mobilized anticipating the worst. Right from the beginning our efforts were incredibly intense."

But how, I persist, could the Coast Guard have said immediately that it didn't seem oil was leaking?

Allen replies that for two days after the rig sank, there was so much silt kicked up that the remote vehicles had a hard time seeing anything. "You've gotta remember, this is a place where there's no human access. Everything has to be done remotely. It took almost seventy-two hours to understand what was going on. Initially there were leaks in three places where the pipe was kinked. It took a few days before the abrasive sand and everything coming out finally ground a larger hole in it."

"You can do Monday-morning quarterbacking," Lubchenco says to me a bit pointedly, "but at the time there was extraordinary effort. We mobilized ships and planes and people, assuming that it would be really bad."

"How," I posit, "is it possible to assume it's really bad when you kept putting out flow estimates that were one-sixtieth of the—"

Allen interrupts: "I was irritated over the early declarations about whether this was one or five thousand barrels. I listened to that and I thought, 'This is crazy—*nobody* knows.'"

Lubchenco wants me to appreciate that "for weeks and weeks, the primary focus was stopping the flow." She adds, "We all thought the early estimates were very low, but we didn't have a way of getting better estimates initially. In every other spill, the amount of oil was estimated from the surface. But this time a lot of oil was staying deep. We were really frustrated early on because BP said, 'No, we don't want anybody else down there.'"

"They were just focused on stopping the oil," says Allen. "That's when it's our job to say, 'Here's our national priorities; here's how we're gonna do it.' With the amount of remotely operated vehicles moving around down there, it is stunning we didn't have a major accident. We had ROVs bump into one other and knock things off; that happened twice, almost with major consequences. It's like the eight near midair plane collisions. There was an unbelievable amount of stuff going on within one square mile; that never happened before. The area where oil was coming to the surface sometimes had thirty-five vessels all in one square mile. But at one point," he adds, "I said, 'We're gonna establish an independent estimate of flow rate.'"

Lubchenco continues: "Thad ordered them to let the group go down and make those measurements."

Allen takes another sip of his coffee and glances at his watch. I know we don't have a ton of time left. He poses the next question himself: "Could we have stopped it sooner? In hindsight, we might have saved two to three weeks. But we were operating with an abundance of caution."

"We had very good reason to believe," Lubchenco informs me, "that the well had been damaged. The concern was, if we built too much pressure in it, the oil would start breaking out through the seafloor. And *then*—it would be completely uncontrollable."

"That's the Armageddon we all feared," Allen adds. "So we had eighty-five days of a different spill coming to the surface in a different way in a different place every day, depending on winds and current conditions. We had a hundred thousand different patches of oil from Louisiana to Florida. Because the oil spill contingency plan didn't call for enough equipment, we were behind the power curve for six or seven weeks. We had three kinds of responses to the oil: skim it, burn it, or disperse it."

I point out that *all* of the oil was coming out of one pipe: "You had your hands around it right there at the source, and you let it get away."

"We just did not have the ability to capture it at the source," Lubchenco replies. "And we didn't have the boats, skimming equipment, right weather—or the kind of oil—to let us just put a ring around it and capture it at the surface. It sounds like that should have been easy," she says. "But it was *not.*"

"When the Deepwater Horizon's pipe broke," Allen affirms, "there was nothing capable of capturing that oil. A system should have been in place, but it wasn't, because we were focused on tanker spills. Oil spill response in this country is based on *Exxon Valdez.*"

"That's *painfully* obvious," Lubchenco concurs.

"The assumption was, 'We'll never have a failed blowout preventer.' "

But we know blowouts happen, I insist. What they finally installed on July 15 to stop the leak, they should have already had in a warehouse before April 20.

"I couldn't agree more," Lubchenco affirms.

"We're not gonna sit here and defend the fact that it wasn't there," Allen says.

Of the kinds of things that *were* there, Lubchenco says that she hadn't anticipated how much their response options depended on weather. "I mean, skimming just doesn't work in rough water."

"But in rough water you can disperse," Allen points out, "and the motion helps mix the dispersant and oil so it's not just laying on top of it."

"Oil is toxic and nasty," Lubchenco says. "There are no choices that are risk-free."

"No right choices," Allen agrees. "Nothing good happens when oil gets into water."

"Skimming and burning were just very ineffective. That wasn't getting us very far," Lubchenco says.

"The Hobson's choice was," Allen says, "accept the fate of the oil in the ocean, or accept the fate of the oil along the shore. There was no easy answer."

"So," Lubchenco says, "the decision was made—not by me; on the advice of our scientific support people—that, on balance, using dispersants was the right thing to do."

But the thing that sticks in people's ribs, I say, is this: dispersed oil is still oil. And it's still in the waters of the Gulf.

"Agreed," says Lubchenco.

"The dispersants don't make the oil go away," Allen acknowledges.

"And that's problematic," Lubchenco points out. "So let me continue. Now, the thing is, dispersed oil is available to be biodegraded much, much faster." That's because the crude's surface area gets vastly increased, facilitating microbial attachment, basically cutting it into little pieces that bacteria can more easily eat. "And that's the *whole* rationale for using chemical dispersants," she says.

"But," she adds, "twice as much was dispersed physically as chemically. When such hot oil, being shot out under pressure, suddenly hits such cold water, it fractionates into microscopic droplets. 'Dispersed' *doesn't mean* it hasn't had impact. I think it has likely had. Early on we got ships out there to get baseline data of things like plankton and bluefin tuna larvae—as much as possible. When this disaster happened, just-spawned shrimp and crabs and fish were in the drifting plankton. The plankton, I think, could have been very seriously affected. Something like eighty to ninety percent of the economically important fish populations in the Gulf depend on the marshes and estuaries for part of their lives; they move back and forth. For them, this could not have come at a worse time. But it's next to impossible to document—so far—what's happened to them."

"So," she sums up, "1.25 million barrels of neutrally buoyant chemically and physically dispersed oil ended up drifting in the water at depths between thirteen hundred to forty-three hundred feet. And *that*—that's not good. I mean, oil is inherently really toxic. We won't know for a while—we really won't know for decades—but it's likely it's had very serious impacts."

I'm surprised that Lubchenco seems to think things will likely turn out worse than I'm betting they will. And that she's saying so. I might have expected her to downplay the long-term effects while I insisted that things were likely worse. How odd that we're not following that script. It's in her best professional interest for the long-term effects of this to be minimal. Because the results won't be in for a long time and some damage may never be known, Jane's got the option of hiding behind uncertainty to make herself look better. She could easily put on a game face and say she thinks the Gulf will bounce right back. That she's not doing this suggests both admirable integrity and a cool hand on the tiller.

"Even if the oil degrades pretty rapidly," Lubchenco says, "and is gone in, like, a year, its impact on populations may already be very, very substantial. I have very grave concerns about impacts this has had for populations that were already depressed and longer-lived species."

"So," adds Allen, "since Kemp's ridley turtles don't start breeding until they're twelve years old, we won't know for twelve years what happened with this year's hatchlings."

Kemp's ridleys were probably the most at-risk of the Gulf's wildlife. "This was a critically endangered population finally coming back because of turtle excluders in fishing nets and better protection of their nesting beaches in Texas and Mexico," Jane notes. "They were really coming back. And then with this, they were hit very, very hard—they were *hammered*."

What about dolphins? "It doesn't look like many died," Lubchenco says, brightening a bit. "We brought some people in to tag them. And from the tagging we can see them still moving around, alive and seeming to behave normally."

Corals? Oyster beds?

"So far we haven't seen oil in corals," Lubchenco says, "though we haven't finished looking everywhere yet. But oysters—they're toast."

"In Louisiana," says Allen, "they decided to send so much fresh water through the delta to keep the oil outside the marshes, they committed oystercide—or whatever you'd call it."

In May, Louisiana's Governor Bobby Jindal, supported by local parish officials, ordered technicians to open giant valves on the Mississippi River, releasing torrents of fresh water with the idea of pushing any oil off the coast. The fresh water largely demolished southeastern Louisiana's oyster beds, killing far more oysters than did the oil.

"Again, that's local officials doing what they think is expedient. Oyster beds have to be reseeded and started over again."

(In September, it seems as if shallow-water coral reefs may escape oil damage. But in November 2010, researchers on one of NOAA's ships will find dead deep-water corals seven miles southwest of the Macondo well, 4,500 feet deep, where researchers last spring had discovered drifting hydrocarbon plumes. The expedition's chief scientist, Charles Fisher of Pennsylvania State University, will comment, "We have never seen anything like this at any of the deep coral sites that we've been to, and we've been to quite a lot." Dr. Lubchenco will say, "This is precisely why we continue to actively monitor the Gulf.")

Lubchenco starts to gather her things and says, "I really have to get going, but we've been talking about the ecological dimensions of this disaster, and yet the human disaster here is very, *very* real."

"We on the government side need to get better at making it known that we understand that people's emotional reactions—the passion and the angst—*are* the right way to feel about this," Allen adds. "This is *grieving*."

"So many communities, so many individuals are just devastated," Lubchenco points out. "The federal government did an insufficient job communicating real compassion. Not that those in government didn't care, but it didn't come across to people as compassion."

"I'll give you a good example of the problem here," the admiral says. "We are not allowed under the Oil Pollution Act of 1990 to use money from the Oil Spill Liability Trust Fund to do anything about behavioral health."

Lubchenco shakes her head and breathes, "Isn't that *just* ridiculous?"

"It's not allowed," Allen says. "Nor was BP, the responsible party, required to do anything for mental wellness."

"A *lot* of changes need to be made," Lubchenco says, though her agency's charge is oceans and weather forecasting, not emotional health.

Of course, that's the space between angst and suicide.

"So we had a charter boat captain in Alabama who did commit suicide," Allen says somberly. "So I went to BP and said, 'You have to give money to these states to set up an 800 number for suicide prevention counseling.' This had nothing to do with my job as national incident commander. But it's what the country expects out of the whole government response. And that's a conversation we have to have."

This conversation is coming to a close. We've been here for two hours; Jane's got a plane to catch. We rise and shake hands, and I thank them for their valuable time.

Allen begins to turn away, then suddenly turns back to me and says, "One last thought: large corporations and the government are usually not capable of the conspiracies that are attributed to them. A lot of what seems like conspiracies has to do with just plain culture, capacity, and ineptness."

Lubchenco nods and agrees: "I can't *believe* all the conspiracy theories I've heard." She rolls her eyes and smiles, and says, "It's been hell."

⁂

On Sunday, September 19, with the cementing done, Admiral Allen declares America's worst oil leak "effectively dead."

Two months after the oil stopped flowing, currents and oil-eating microbes continue steadily dissipating and degrading the oil. NOAA's Dr. Steven Murawski says that even to the most sensitive instruments, the oil deep in the open Gulf is fading to levels barely detectable, making the underwater plumes "harder and harder to find." The oil is becoming, Murawski says, "like a shadow out there."

In the last week of September and the first week of October, NOAA will reopen fishing in almost 17,000 square miles of Gulf federal waters. BP ends its boat employment, and Jane Lubchenco's agency—in an unusual move—opens fishing for the hugely popular red snapper as a boost to tourism and to anglers who've missed a whole spring and summer's fishing. If fish could think, they might find themselves feeling nostalgic for the summer of *pax petroleum.*

Those fish that got a few months' peace—oil notwithstanding—remain to be caught in numbers anglers are not accustomed to. "Red snapper are unbelievable right now," says one fisherman. "You could put a rock on the end of a string and they'll bite it."

What are scientists finding?

"It's not what you would have guessed," says Dauphin Island Sea Lab senior marine scientist John Valentine. "You would have expected something horrible, but that's not what we're seeing." Ironically, the blowout's most powerful environmental effect seems to be both indirect and positive: the fishing closures. Fishing is designed to kill things living in the sea, and it does so effectively. It's been by far the major agent of change in the world's oceans until now. And if all of us use petroleum, most of us also eat seafood. Valentine's finding about three times as many fish now compared to before the blowout, and they're bigger.

Likewise, Sean Powers of the University of South Alabama finds that even some coastal Alabama shark populations have tripled in number, and a lot of that comes from this year's young. Normally, shrimp nets kill a lot of small sharks. "It's just been amazing how many more sharks we are seeing this year," he says. "I didn't believe it at first. What's interesting to me," he adds, "we are seeing it across the whole range, from the shrimp all the way up to the large sharks." Ken Heck, also of the Dauphin Island Sea Lab, says species whose young live in sea-grass meadows also appeared robust. "It's quite amazing. Everyone speculated that an entire year class of larvae and young might have been lost, smothered in oil," Heck says. "That hasn't happened."

Oddly, this seems almost troubling to researchers wishing to understand the effects of oil itself. "The problem with the fishing closure is, that impact is so large it is probably going to swamp any impact of the oil spill," Powers says. "We're not saying we didn't lose any fish to the spill. We're saying it is going to be harder to detect any smaller changes due to oil spill contamination. We'll have to look carefully."

Valentine notes, "This was the first time we've ever seen such a large-scale cessation of fishing." I note, nature's resilience is truly magnificent.

Well, what does all that tell us? It tells me that restoring a healthy ocean must also involve mixing fishing into the big-picture strategy. An oil lease is a piece of seafloor specially set aside for oil extraction. Why don't we also set aside other places, to restore a better ocean? If it's in the national interest to produce oil and gas, it's as much in the national interest to protect other social and economic and living interests. There are more than eight thousand active oil and gas leases on America's outer continental shelf. In addition to setting those places aside for taking oil and gas, why don't we insist on setting special places aside for protection? Right now, less than 1 percent of all federal waters of the United States have been protected as national marine sanctuaries or any other safeguarded areas.

In an e-mail to various folks, the indefatigable Dr. Sylvia Earle exhorts:

> To compensate the Gulf, and provide hope for recovery, actions should be taken ASAP to identify and protect areas that are still in good shape. Obama has the power to do as Presidents T. Roosevelt and G. W. Bush have done in the past—under the Antiquities Act—to declare National Monuments. What better way to "give back to the Gulf," and to the people whose livelihoods depend on a healthy Gulf, than to protect the deep reefs and string of "topographic highs" in the Northern Gulf, the spawning areas for tuna, the critical places for menhaden, grouper, snapper, shrimp and others, as well as the vital—but neglected—seagrass meadows of Florida's Big Bend area. Respect for the importance of the floating forests of sargassum and their role in providing nursery areas, food and shelter—as well as taking up carbon and generating oxygen—might be considered in an overall recovery plan. This could be the moment to act to secure protection for Pulley Ridge, the extraordinary system of deep reefs 150 miles offshore from Sarasota. And time is of the essence. Maybe now is the time, when the need is so obvious, to act.

When BP announces it has spent $9.5 billion cleaning up its mess (a helium-filled figure that will keep floating upward over the next weeks and months), I'm not sure how I feel about it. It's nice of BP, but they could have saved themselves and all of us the trouble by saying, "Wait a minute" when they saw pressure on the gauge during the afternoon of April 20. Maybe next time they won't keep driving when the oil light comes on.

But what will the next time be? Just as the federal response plan failed this blowout because it was designed to fight the next *Exxon Valdez,* we're gonna need something that doesn't just plan to fight the last war. Some added thought is called for. Some vision. What might be the next big problem in the tapping and transport of oil, the one nobody is anticipating?

More than 27,000 abandoned wells lurk beneath the Gulf of Mexico. With some abandoned as long ago as the 1940s, deteriorating sealing jobs may already be failing. About 13 percent of them are categorized in government records as "temporarily abandoned." Temporarily abandoned wells are supposed to be plugged within a year. That's routinely ignored. Many "temporarily abandoned" wells have been sitting since the 1950s and '60s. No one—not the oil companies, not the government—is checking whether they are leaking. Meanwhile, cement and pipes never intended to last so many years are aging in the seafloor. All this raises the possibility that old wells may spontaneously blow out.

If that happens, it seems unlikely that companies will rush forward to accept responsibility. BP alone has abandoned about 600 wells in the Gulf. "It's in everybody's interest to do it right," says a spokesman for Apache Corporation, which has abandoned at least 2,100 wells in the Gulf. But early on, the rules were less strict than they are today, and many—tens of thousands—of wells are poorly sealed.

Texas alone has plugged more than 21,000 abandoned wells to control pollution in state waters. Other places have similar problems. California has resealed scores of abandoned wells. But in deeper federal waters, the U.S. Minerals Management Service has typically inspected only the paperwork, not the real job. Over five years, from 2003 through 2007, the MMS fined seven companies a total of only $440,000 for improper plug-and-abandonment work.

The Government Accountability Office, which does congressional investigations, warned as early as 1994 that leaks from offshore abandoned wells could cause an "environmental disaster." The GAO pressed for inspections of abandonment jobs, but nothing was done. A 2001 Minerals Management Service study noted concern that "some abandoned oil wells in the Gulf may be leaking crude oil."

Nothing came of that, either. In 2006 the U.S. Environmental Protection Agency said, "Well abandonment and plugging have generally not been properly planned, designed and executed." The EPA was talking about wells *on land*, where wells abandoned in recent decades have leaked so routinely that dealing with the problem is called "replugging" or "reabandonment." The GAO estimated that 17 percent of the nation's land wells had been improperly plugged. If offshore wells are no worse, thousands of wells in the Gulf of Mexico alone are badly plugged.

Various people, from the president on out, have called this blowout "the worst environmental catastrophe in American history." Some simply said "in history."

Well, no. I heard a lot of catastrophizing on both ends. I read everything from how the Deepwater Gulf blowout would trigger a massive seafloor methane release that would kill the whole world (actually, global warming is beginning to trigger a massive methane release), to BP's Tony Hayward saying it's nothing much, really. Usually the truth is somewhere in the middle. But this time, the truth is closer to one end of those extremes.

BP's Tony Hayward became the most hated man in America for saying that the amount of oil leaked was "tiny" compared with the "very big ocean." He might have been making excuses for BP; I'm certainly not. BP, Transocean, Halliburton, and others had to screw up big-time half a dozen different ways to get this well to blow. And in order to be 100 percent unprepared, Big Oil and our own antigovernment government regulators had to ignore Ixtoc and the world's other blowouts, cut and paste walruses into their response plans, and do the sloppiest, most cynical job money can buy. BP, Transocean, and Halliburton's legal, fiscal, and public relations nightmare is far from over, and that's fine by me. But whether he realized it or not, in one sense Hayward was right.

In the blowout, 206 million gallons of oil mixed with the Gulf's 660 quadrillion gallons of water. That volume of water could greatly dilute the oil. But the carbon dioxide we're adding to the atmosphere isn't getting diluted; it's building up.

The oil that is getting into the ocean has everyone's attention. It was supposed to be refined to help power civilization, not spew waste and devastation. But Plan A, *burning* the oil—and coal, and gas—in our engines is continually adding carbon dioxide to the atmosphere at the inconceivable rate of a thousand tons a second, billions of tons a year. *That* spill is invisible. Rather than washing up on one coast and scaring tourists, this spill, spread in time and space, is slowly coating the whole world. There is no single company at which to point fingers.

In this, we are all involved. Our everyday use of fossil fuels is changing the atmosphere, ruining the world's oceans. The top-tier journal *Science* has published a special issue called "Changing Oceans" that summarizes major changes being seen in marine life and ocean function—and the significant implications for human health and the food supply. Calling the world's oceans the heart and the lungs of the planet, one of the authors says, "It's as if the Earth has been smoking two packs of cigarettes a day." The oceans produce about 50 percent of the oxygen we breathe, but absorb 30 percent of the carbon dioxide our burning produces, and also absorb more than 85 percent of the extra heat trapped by carbon dioxide and other greenhouse gases. Carbon dioxide isn't just warming the world; it's making seawater more acidic.

Because we've bet the house on burning oil, coal, and gas, our atmosphere's concentration of carbon dioxide is a third higher now than at the start of the Industrial Revolution. As environmental catastrophes go, that dwarfs the Gulf blowout of 2010. That's just a fact, and, again, I state it with no intended disrespect to the people of the Gulf whose lives were so horrendously distorted by the blowout.

The worst environmental disaster in history isn't the oil that got away. The real catastrophe is the oil we *don't* spill. It's the oil we run through our engines as intended. It's the oil we burn, the coal we burn, the gas we burn. The worst spill—the real catastrophe—is the carbon dioxide we spill out of our tailpipes and smokestacks every second of every day, year upon decade. *That* spill is changing the atmosphere, changing the world's climate, altering the heat balance of the whole planet, destroying the world's polar systems, killing the wildlife of icy seas, killing the tropics' coral reefs, raising the level of the sea, turning the oceans acidic, and dissolving shellfish. And as the reefs dissolve and the productivity of oceans and agriculture destabilizes, so will go the food security of hundreds of millions of people.

Fossil fuels are great fuels. They are very energy-dense for their size and weight. You can't eat them, so their use as fuel doesn't compete with people's needs for food or farmland. (If the entire U.S. corn crop went to make ethanol, it would replace only 15 percent of America's gasoline usage.) Clean renewables have drawbacks. A coal plant can keep producing energy whether or not the sun is shining or the wind is blowing. Wind is clean, but many people object to the sight of turbines (they should see how the oil rigs dominate the view in the Gulf), to their sound, or to the fact that they kill birds and bats. Large-scale solar projects need a lot of land.

But fossil fuels have drawbacks, too. One, they're not eternal and won't last forever. Their production is likely to peak in a few decades. Two, in the process of getting them, workers die and dictators thrive. Three, they're hurting the world's life-support capacity.

It is hard to imagine how solar power or wind or algae could power all of civilization. But not so very long ago, the present scale and difficulty of coal mining, oil drilling, and civilization itself—for that matter—was impossible to envision. Even though fossil fuels elicited giddy attraction, ramping them up to dominance took the better part of a century. No matter how good a fuel is, it takes time to create the technology to produce it, the infrastructure to transport it, and the consumer demand for it.

We have to do something. The Gulf of Mexico accounts for almost a third of America's oil production and most new discoveries. Most of the world's land has been sucked dry, especially in the United States. And elsewhere, it's being put out of reach by autocratic, self-protecting governments.

Many people say—*and they have a point*—that if Americans do not want to hand even more money and clout to the likes of Iran, Saudi Arabia, Russia, and Venezuela (I certainly don't), we should drill more at home. I would add: and America should harness *all* our domestic sources of energy. We should get on an emergency war-footing crash program for creating the jobs and building the infrastructure to surpass China and northern Europe's renewable-energy race, summon the determination to lead the world into the eternal-energy economy, and emerge again as the greatest country on Earth. You'll hear me say this again, because, really, that's the Big Picture here.

In America, there hadn't been a big offshore oil well leak in forty years. Underwater pipeline leaks declined from an average of 2.5 million gallons per year in the early 1980s to just 12,000 gallons a year in the early '00s, according to the Congressional Research Service. The National Research Council estimates that offshore drilling, tankers, and pipelines account for 5 percent of the oil that gets into in U.S. waters, while shipping accounts for 33 percent, and 62 percent comes from natural seeps (natural seeps send perhaps 1,000 to 2,000 barrels of oil into the northern Gulf daily, though estimates vary a lot).

This leads us to a question. What's the worst part of America's oil addiction: funding foreign dictators, warming the air and acidifying the seas, or drilling holes? Perhaps we should not, after all, keep America's waters closed to our thirst for oil. Perhaps we should just drink a wider variety of energies that are better for us, better for everything. Americans pay a fraction of the full cost of a gallon of gasoline, if you count the costs of pollution and wars to maintain access. Not at the pump, anyway.

The best way to respond to the Gulf disaster? Not washing oil off birds, picking up turtles, spraying dispersants, or cleaning beaches. Rather, pulling the subsidies out from under Big Petroleum. Since we pay those subsidies in our income taxes and lose sight of them, it'd be better to put them right in our gasoline and oil taxes and let ourselves be shocked at the pump by the true cost we're paying—and hurry toward better options.

We have a shortage of time and a long way to go. Considering how much damage carbon dioxide is doing to our atmosphere and ocean, it would be unwise to simply assume we won't need nuclear energy. We in the United States already get about a fifth of our electricity from nuclear, and nuclear plants provide similar proportions for several other countries. (In one sense, all our energy is nuclear, since the sun itself derives its energy from hot nuclear fusion reactions similar to those in a hydrogen bomb.) It might be necessary to accept the drawbacks of more nuclear as a bridge to cleaner energy. But the dangers with nuclear's spent fuels and potential weapon-building materials make nuclear hard to swallow. The economics are difficult, too: construction and decommissioning costs remain quite high, likely too high. Things can go wrong with nuclear in ways that simply can't happen with other zero-carbon-emissions sources. (Terrorists can't do much with windmills and solar panels.) Reasonable people can reach different conclusions about nuclear energy. My conclusion is that our attention ought be firmly focused on clean renewables.

Solar energy delivers power enough to meet our projected future energy demands all by itself. But today solar electricity is just 0.1 percent of total world electricity production and solar heating, a similar 0.1 percent of world energy production. For wind to generate 20 percent of U.S. electricity, we'd need almost ten times the capacity we currently have and we'd need to build 100,000 new turbines.

Replacing just half the power generated today by oil, coal, and gas would require 6 terawatts; renewable energy sources now generate only half a terawatt. But transitions have always been forced by shortages. Necessity is the mother of innovation. During the War of 1812, wood shortages around Philadelphia prompted residents to experiment with burning coal. When Edwin Drake drilled the first oil well in the United States, in 1858, whale oil was already getting harder to come by.

One problem with clean fuels is the perception that "they don't offer new services; they just cost more," as one analyst has said. Wrong on both counts, but the statement reveals our inability to understand the effects and costs of energy. The new services are the elimination of toxic pollution, risk, and the climate change and ocean chemistry change that fossil fuels cause. The costs of those things are enormous. The fact that the costs are not in the fuels' at-pump price is a failure of our economics, not a drawback of clean fuels. We are already paying, and we will pay enormous future costs for the effects of climate change on agriculture, coastal cities, coral reefs and fisheries, security, and the abundance and diversity of wildlife worldwide.

Those costs are serious. A moral and practical answer is to engineer the transition. Shifting massive and guaranteed subsidies away from fossil fuels and into clean renewables would be a big part of the way to accomplish what's needed.

But because we don't understand the difference between price, cost, and value, we can't seem to get our minds unstuck and move beyond or around the idea that we simply don't want to "pay more" for better energy. On June 20, 2010, at the height of the agony in the Gulf, a CBS News / *New York Times* poll found that 90 percent of people agreed that "U.S. energy policy either needs fundamental changes or to be completely rebuilt" but that just under half of them—49 percent of people responding—supported new taxes on gasoline to fund new and renewable energy sources. If people don't want to pay more at the pump, the best thing we can do to save money *and* buy time until we have better options would be to conserve energy and improve efficiency. Who are the losers there? Everybody wins. But boosting efficiency, too, runs into political trouble.

If we could build the infrastructure to capture and transport energy from renewable sources, the energy itself—sunlight, wind, tides, geothermal heat—would come for free. That's what "clean, eternal" means. But we hear that energy that comes for free is too expensive. Who tells us clean, eternal energy is too expensive and that it would "wreck the economy"? Why, it's none other than the big brothers: Big Oil and Big Coal! We've crossed that bridge before. There was another time when people vehemently insisted that changing America's main source of energy would wreck the economy. The cheapest energy that has ever powered America was slavery. Energy is always a moral issue.

Big Oil and Big Coal maintain our addiction to their elixirs. But we allow it, rather than freeing ourselves to a more diverse, decentralized, cleaner, stable array of fuel sources.

The real tragedy is that for thirty years we've known that for reasons of national security and patriotism alone, we need to phase out our dependence on oil, coal, and gas. Our foreign dependence, the jobs we're missing out on, the pollution, the worker-killing explosions, the way we enrich dictators and terrorists—we've known all that since the politically induced oil shocks and gasoline lines of the 1970s. I remember those lines with some fondness, because I waited on them as a young high school buck and proud new owner of a used hippie van. But those lines were all we needed to learn that security for the United States, and the globe, requires a future largely free of fossil fuels.

And since the late 1980s, we've known that fossil fuels are also destabilizing the world climate. But Big Oil and Big Coal, using the subsidies we pay them, maintain the weakest government money can buy. So most Republicans and a few indebted Democrats scoff at energy efficiency.

Multinational corporations are by definition not patriotic; they can't afford to be. But we can't afford for them not to be. Their interests are not our interests. For the main reason behind America's decline—in manufacturing, jobs, technological innovation, and moral leadership—we need look no further. Multinational corporations have strangled innovation in its crib. Killed all our first-born ideas and sent the entrepreneurs who could have saved us fleeing to places like China.

China understands its moment. Today China is rapidly becoming the world's leader in wind and solar energy, electric cars, and high-speed rail. It's also the world's greatest lender of money to the United States; we have yoked ourselves to interest payments to the world's biggest totalitarian government, while forking over union jobs, technological leadership, and the American Dream. It's been said that empires are not destroyed from the outside; they commit suicide.

Every president since Nixon has talked about our need to kick our oil addiction. Some were serious. But by failing to sweep money out of politics, we disenfranchise ourselves from control of our own government, our own country.

And yet we simply won't be able to maintain civilization by digging fossil energy out of the ground and burning it. There's not enough. By the middle of this century, well within the life span of many people already born, we are scheduled to add to the world another two billion people—nearly another China plus another India's worth of people. Of the truly great human-caused environmental catastrophes, foremost is the human population explosion. The forests, the fishes, and fresh water are collapsing under the weight of the number of people on Earth right now. All the other global environmental, justice, human development, energy, and security problems either start with or are made worse by the sheer crush of our numbers. The projected growth will squash human potential as billions of poor get poorer, while flaring tensions and igniting violence. Being concerned about overpopulation isn't anti-people; not being worried about it is anti-people.

We run civilization mainly on the energy of long-ago sunlight, locked away in oil and coal. It's time we step into the sunlight itself and phase in an energy future based on harnessing the eternal energies that actually run the planet. Whoever builds that new energy future will own the future. And the nation that owns the energy future will sell it to everyone else. I'd rather that nation be the United States of America. Did we really wage a decades-long Cold War just so we could anoint China the world's Big Box Store? So we could be indebted to China for generations? Did we really hand world leadership to the largest autocratic nondemocracy in the history of the world because all it can offer is low wages, no unions, and cheap goods? Is that all it takes to secure our surrender? Where are the patriots?

New details about the Gulf blowout of 2010 will continue to bubble up for quite some time. As the birds of autumn begin rowing through the air near my Long Island home, I begin seeing gleaming white gannets on their way south after nesting in Canada. Last spring, the first oiled bird whose image made news was a gannet that had spent its winter in the Gulf. I realized that the bird would never return to its Canadian nesting grounds, meaning the oil would create problems reaching far beyond the Gulf of Mexico. Juvenile gannets can spend several years in the Gulf, and now I get word that Suncoast Seabird Sanctuary, near

Tampa, is still nursing about eighty gannets it received during the summer, debilitated by oil but alive. That means hundreds of gannets died. In Canada, seabird biologist Bill Montevecchi says that his tagging studies suggest that the oil posed a threat to as many as a third of Canada's gannets. Even if all goes well from now on, it will take a few years for the population, nesting well over two thousand miles from the Gulf, to replace those birds lost to the Gulf oil slicks. And that's just gannets.

The blowout is both an acute tragedy and a broad metaphor for a country operating sloppily, waving away risks and warnings, a country that does not use care in stewarding its precious gifts, a country concerned only about the next little while, not the longer time frames of our lives or our children's futures.

In the meantime, we are left with the Gulf itself. Regardless of how fast the Gulf's waters, wildlife, and wetlands recover from this blowout, Gulf residents will be left with scars and years-long pain. The stamp of this will be on lives, families, marriages, and children for quite a while to come.

During the last week of September, "moderate to heavy oil" washes onto almost one hundred miles of Louisiana's shoreline. And when Louisiana State University scuba divers go for a look into the waters off of St. Bernard and Plaquemines Parishes, they see "oil everywhere."

REFERENCES

BLOWOUT!

4 *Deepwater Horizon size, and insured for over half a billion dollars* S. Mufson, "Gulf of Mexico Oil Spill Creates Environmental and Political Dilemmas," *Washington Post,* April 27, 2010. *See also* "Deepwater Horizon," at Wikipedia.org; accessed on November 12, 2010.

4 *No serious injuries in seven years* "Blowout: The Deepwater Horizon Disaster," CBS News, May 16, 2010; http://www.cbsnews.com/stories/2010/05/16/60minutes/main6490197.shtml.

5 *Drilling progressed to about 4,000 feet* "Oil Spill Postmortem: BP Used Less-reliable, but Cheaper Drilling Method on Deepwater Horizon," McClatchy-Tribune News Service, May 23, 2010.

5 *$58 million over budget* "AFE Summary for the Macondo Well," http://www.deepwaterinvestigation.com/external/content/document/3043/914919/1/AFE%20Summary%20for%20the%20Macondo%20Well.pdf.

7 *Casing and drill pipe and "making a trip* M. Raymond and William Leffler, *Oil and Gas Production in Nontechnical Language* (Tulsa, OK: PennWell Corp., 2006), p. 104.

9 *It was a world-class rig* "Deepwater Horizon's Cementing Plan Is Under Scrutiny." The well-drilling description is based on schematic and notes from the *Times-Picayune,* 2010; see schematic at http://media.nola.com/2010_gulf_oil_spill/photo/oil-halliburton-cement-052010jpg-e618a2271a66c847.jpg.

9 *The rig was getting dated, and Lloyd's findings* R. Brown, "Oil Rig's Siren Was Kept Silent, Technician Says," *New York Times,* July 23, 2010, p. A1; http://www.nytimes.com/2010/07/24/us/24hearings.html?_r=1&hp.

10 *The rig lost all its fluid* I. Urbina, "BP Used Riskier Method to Seal Well Before Blast," *New York Times*, May 26, 2010, p. A1; http://www.nytimes.com/2010/05/27/us/27rig.html?_r=2 .

10 *"We got to a depth of 18,260 feet"* "USCG/BOEM Marine Board of Investigation into the Marine Casualty, Explosion, Fire, Pollution, and Sinking of Mobile Offshore Drilling Unit Deepwater Horizon, with Loss of Life in the Gulf of Mexico," April 21–22, 2010," testimony of John Guide, transcript p. 87, lines 1–25, July 22, 2010; http://www.deepwaterinvestigation.com/go/doc/3043/856503/.

11 *"While drilling that hole section we lost over 3,000 barrels of mud"* "USCG/MMS Marine Board of Investigation into the Marine Casualty, Explosion, Fire, Pollution, and Sinking of Mobile Offshore Drilling Unit Deepwater Horizon, with Loss of Life in the Gulf of Mexico 21–22 April 2010," testimony of Mark Hafle, transcript pp. 42–43, lines 1–25, 1–5, May 28, 2010;http://www.deepwaterinvestigation.com/go/doc/3043/670171/ .

11 *At over $250 per barrel* P. Parsons, "The Macondo Well: Part 3 in a Series About the Macondo Well (Deepwater Horizon) Blowout," *Energy Training Resources, LLC*, July 2010, pp. 17, 15; http://www.energytrainingresources.com/data/default/content/Macondo.pdf. *See also* M. J. Derrick, "Cost-Effective Environmental Compliance in the Gulf of Mexico," *World Energy Magazine* 4, No. 1 (2001); http://www.derrickequipment.com/Images/Documents/wemdvol4no1.pdf.

11 *Delays cost a week and led to a budget add-on of $27 million* "AFE Summary for the Macondo Well," http://www.deepwaterinvestigation.com/go/doc/3043/914919.

11 *Forty-three days behind* R. Brown, "Oil Rig's Siren Was Kept Silent, Technician Says," *New York Times*, July 23, 2010, p. A1; http://www.nytimes.com/2010/07/24/us/24hearings.html?_r=1&hp.

11 *Crew is informed they'd lost $25 million,* "Blowout: The Deepwater Horizon Disaster," a *60 Minutes* report that features Mike Williams, CBS News, May 16, 2010; http://www.cbsnews.com/stories/2010/05/16/60minutes/main6490197.shtml.

11 *The final section of the well bore extended to 18,360 feet* U.S. Congress, House Committee on Energy and Commerce, June 14, 2010, letter from Henry Waxman and Bart Stupak to Tony Hayward, 111th Congress: p. 4; http://energycommerce.house.gov/documents/20100614/Hayward.BP.2010.6.14.pdf.

12 *"Cost is not a deciding factor"* "USCG/MMS Marine Board of Investigation into the Marine Casualty, Explosion, Fire, Pollution, and Sinking of Mobile Offshore Drilling Unit Deepwater Horizon, with Loss of Life in the Gulf of Mexico," April 21–22, 2010, testimony of David Sims, May 26, 2010; http://www.deepwaterinvestigation.com/go/doc/3043/670067/.

12 *"a win-win situation"* USCG/BOEM Marine Board of Investigation into the Marine Casualty, Explosion, Fire, Pollution, and Sinking of Mobile Offshore Drilling Unit Deepwater Horizon, with Loss of Life in the Gulf of Mexico," April 21–22, 2010, testimony of John Guide, transcript p. 397, lines 18–23, July 22, 2010; http://www.deepwaterinvestigation.com/go/doc/3043/856503/.

12 *"With every decision, didn't BP reduce"* "Video: Deepwater Horizon Joint Investigation Footage," July 22, 2010, Part 36, posted on July 24, 2010; http://www.dvidshub.net.

13 *"without a doubt a riskier way to go"* I. Urbina, "BP Used Riskier Method to Seal Well Before Blast," *New York Times*, May 26, 2010, p. A1; http://www.nytimes.com/2010/05/27/us/27rig.html?_r=2.

14 *Jason Anderson's concerns over safety* "BP Toolpusher Jason Anderson Feared for His Life," June 3, 2010; http://seminal.firedoglake.com/diary/52541\. *See also* "Deepwater Horizon Victim—Jason Anderson," http://www.awesomestories.com/assets/deepwater-horizon-victim-jason-anderson. *See also* R. Arnold, "Son Warned of Trouble on Doomed Rig," August 16, 2010; http://www.click2houston.com/news/24652197/detail.html. *See also* T. Junod, "Eleven Lives," *Esquire*, August 16, 2010; http://www.esquire.com/features/gulf-oil-spill-lives-0910. *See also* J. Carroll and Laural Brubaker Calkins, "BP Pressured Rig Worker to Hurry Before Disaster, Father Says," Bloomberg, May 27, 2010; http://www.bloomberg.com/news/2010-05-28/bp-pressured-oil-rig-worker-to-hurry-before-fatal-gulf-blast-father-says.html.

15 *"His experience was largely in land drilling"* B. Casselman and Russell Gold, "BP Decisions Set Stage for Disaster," *Wall Street Journal*, May 27, 2010; http://online.wsj.com/article/SB10001424052748704026204575266560930780190.html?KEYWORDS=negative+test.

15 *"I raised my concerns"* "USCG/BOEM Marine Board of Investigation into the Marine Casualty, Explosion, Fire, Pollution, and Sinking of Mobile Offshore Drilling Unit Deepwater Horizon, with Loss of Life in the Gulf of Mexico," April 21–22, 2010," testimony of Paul Johnson, transcript p. 205, lines 7–17, August 23, 2010; http://www.deepwaterinvestigation.com/external/content/document/3043/903575/1/USCGHEARING%2023_Aug_10.pdf.

15 *"Mr. Bob Kaluza called me to his office"* "USCG/BOEM Marine Board of Investigation into the Marine Casualty, Explosion, Fire, Pollution, and Sinking of Mobile Offshore Drilling Unit Deepwater Horizon, with Loss of Life in the Gulf of Mexico," April 21–22, 2010, testimony of Leo Lindner, transcript pp. 272 and 313, July 19, 2010; http://www.deepwaterinvestigation.com/go/doc/3043/856483/.

16 *"Let's face it"* W. Semple, e-mail to Mark Loehr, September 19, 2010.

16 *"We're in the exploration group"* "USCG/BOEM Marine Board of Investigation into the Marine Casualty, Explosion, Fire, Pollution, and Sinking of Mobile Offshore Drilling Unit Deepwater Horizon, with Loss of Life in the Gulf of Mexico," April 21–22, 2010, testimony of Ron Sepulvado, transcript p. 63, lines 5–6, July 20, 2010; http://www.deepwaterinvestigation.com/go/doc/3043/856499/.

16 *Production zone between 18,051 and 18,223 feet* BP's Deepwater Horizon Accident Investigation Report, September 8, 2010; http://www.bp.com/liveassets/bp_internet/globalbp/globalbp_uk_english/incident_response/STAGING/local_assets/downloads_pdfs/Deepwater_Horizon_Accident_Investigation_Report.pdf.

17 *"The biggest risk associated with this cement job"* "USCG/BOEM Marine Board of Investigation into the Marine Casualty, Explosion, Fire, Pollution, and Sinking of Mobile Offshore Drilling Unit Deepwater Horizon, with Loss of Life in the Gulf of Mexico," April 21–22, 2010, testimony of John Guide, transcript p. 87, lines 2–4, July 22, 2010; http://www.deepwaterinvestigation.com/go/doc/3043/856503/.

17 *Nitrified foamed cement* P. Parsons, "The Macondo Well: Part 3 in a Series About the Macondo Well (Deepwater Horizon) Blowout," *Energy Training Resources, LLC*, July 15, 2010; https://www.energytrainingresources.com/data/default/content/Macondo.pdf.

17 *The depth created concern* D. Hammer, "Deepwater Horizon's Ill-Fated Oil Well Could Have Been Handled More Carefully, Hearings Reveal," May 29, 2010, *NOLA.com;* http://www.nola.com/news/gulf-oil-spill/index.ssf/2010/05/hammer-hearings-cg.html.

17 *"That nitrogen, it could be a bad thing"* "USCG/MMS Marine Board of Investigation into the Marine Casualty, Explosion, Fire, Pollution, and Sinking of Mobile Offshore Drilling Unit Deepwater Horizon, with Loss of Life in the Gulf of Mexico," April 21–22, 2010, testimony of Jimmy Harrell, transcript p. 72, lines 20–22, May 27, 2010; http://www.deepwaterinvestigation.com/go/doc/3043/670139/.

18 *"We were concerned"* "USCG/MMS Marine Board of Investigation into the Marine Casualty, Explosion, Fire, Pollution, and Sinking of Mobile Offshore Drilling Unit Deepwater Horizon, with Loss of Life in the Gulf of Mexico," April 21–22, 2010, testimony of Mark Hafle, transcript p. 45, lines 8–12, May 28, 2010; http://www.deepwaterinvestigation.com/go/doc/3043/670171/.

18 *BP held an in-house contest* A. G. Breed, "The Well Is Dead, but Gulf Challenges Live On," Associated Press, September 19, 2010.

19 *"To see if adding more centralizers will help," and related discussion* U.S. Congress, House Committee on Energy and Commerce, June 14, 2010, letter from Henry Waxman and Bart Stupak to Tony Hayward, 111th Congress, p. 7; http://energycommerce.house.gov/documents/20100614/Hayward.BP.2010.6.14.pdf.

20 *"It was a bigger risk to run the wrong centralizers"* "USCG/BOEM Marine Board of Investigation into the Marine Casualty, Explosion, Fire, Pollution, and Sinking of Mobile Offshore Drilling Unit Deepwater Horizon, with Loss of Life in the Gulf of Mexico," April 21–22, 2010, testimony of John Guide, transcript p. 307, lines 1–25, July 22, 2010; http://www.deepwaterinvestigation.com/go/doc/3043/856503/.

20 *"The BP Macondo team erroneously believed"* BP's Deepwater Horizon Accident Investigation Report, p. 35, September 8, 2010; http://www.bp.com/liveassets/bp_internet/globalbp/globalbp_uk_english/incident_response/STAGING/local_assets/downloads_pdfs/Deepwater_Horizon_Accident_Investigation_Report.pdf.

20 *"That subject never came up"* "USCG/BOEM Marine Board of Investigation into the Marine Casualty, Explosion, Fire, Pollution, and Sinking of Mobile Offshore Drilling Unit Deepwater Horizon, with Loss of Life in the Gulf of Mexico," April 21–22, 2010, testimony of John Guide, transcript p. 307, lines 1–25, July 22, 2010; http://www.deepwaterinvestigation.com/go/doc/3043/856503/.

20 *"Well, we didn't know if we could find them"* "USCG/BOEM Marine Board of Investigation into the Marine Casualty, Explosion, Fire, Pollution, and Sinking of Mobile Offshore Drilling Unit Deepwater Horizon, with Loss of Life in the Gulf of Mexico," April 21–22, 2010, testimony of John Guide, transcript p. 205, lines 1–25, July 22, 2010; http://www.deepwaterinvestigation.com/go/doc/3043/856503/.

20 *"I don't know of any"* "USCG/BOEM Marine Board of Investigation into the Marine Casualty, Explosion, Fire, Pollution, and Sinking of Mobile Offshore Drilling Unit Deepwater Horizon, with Loss of Life in the Gulf of Mexico," April 21–22, 2010, testimony of John Guide, transcript p. 356, line 22, July 22, 2010; http://www.deepwaterinvestigation.com/go/doc/3043/856503/.

21 *"I never knew it was part of the report"* "USCG/BOEM Marine Board of Investigation into the Marine Casualty, Explosion, Fire, Pollution, and Sinking of Mobile Offshore Drilling Unit Deepwater Horizon, with Loss of Life in the Gulf of Mexico," April 21–22, 2010, testimony of John Guide, transcript p. 269, lines 7–8, July 22 2010; http://www.deepwaterinvestigation.com/go/doc/3043/856503/.

21 *"Although the decision not to use twenty-one centralizers"* BP's Deepwater Horizon Accident Investigation Report, p. 35, September 8, 2010; http://www.bp.com/liveassets/bp_internet/globalbp/globalbp_uk_english/incident_response/STAGING/local_assets/downloads_pdfs/Deepwater_Horizon_Accident_Investigation_Report.pdf.

22 *Kill pill as spacer, and quotes about using it* D. Hilzenrath, "Credibility of BP Oil Spill Study Is Challenged," *Washington Post,* September 11, 2010; http://www.washingtonpost.com/wpdyn/content/article/2010/09/11/AR2010091104579.html.

23 *Spacer fluid abnormally used, and that spacer could have affected the blowout preventer, and "would have required disposal," and "snot on the deck"* D. Hammer, "Finger-Pointing over Deepwater Horizon Explosion Grows Heated," July 19, 2010, NOLA.com; http://www.nola.com/news/gulf-oil-spill/index.ssf/2010/07/finger-pointing_over_deepwater.html. *See also* M. Kunzelman, "Federal Hearings Resume in Oil Spill Probe," Chron.com, July 19, 2010; http://www.chron.com/disp/story.mpl/ap/tx/7115149.html.

23 *"To my knowledge—well, it filled a function"* "USCG/MMS Marine Board of Investigation into the Marine Casualty, Explosion, Fire, Pollution, and Sinking of

Mobile Offshore Drilling Unit Deepwater Horizon, with Loss of Life in the Gulf of Mexico," April 21–22, 2010, testimony of Leo Lindner, transcript p. 321, lines 12–15, July 19, 2010; http://www.deepwaterinvestigation.com/go/doc/3043/856483/.

23 *"Why is all this snot on the deck?"* "USCG/MMS Marine Board of Investigation into the Marine Casualty, Explosion, Fire, Pollution, and Sinking of Mobile Offshore Drilling Unit Deepwater Horizon, with Loss of Life in the Gulf of Mexico," April 21–22, 2010, testimony of Steve Bertone, transcript p. 41, lines 1–4, July 19, 2010; http://www.deepwaterinvestigation.com/go/doc/3043/856483/.

24 *Douglas Brown's statements and disparity about whether Vidrine or Kaluza said they'd do it a certain way* J. Harkinson, "The Rig's on Fire! I Told You This Was Gonna Happen!" *Mother Jones*, June 7, 2010; http://motherjones.com/blue-marble/2010/06/rigs-fire-i-told-you-was-gonna-happen.

24 *"I didn't have no doubts about it"* "USCG/MMS Marine Board of Investigation into the Marine Casualty, Explosion, Fire, Pollution, and Sinking of Mobile Offshore Drilling Unit Deepwater Horizon, with Loss of Life in the Gulf of Mexico," April 21–22, 2010, testimony of Jimmy Harrell, transcript pp. 57–58 and 74. He later says he doesn't recall the remark: transcript p. 80, May 27, 2010; http://www.deepwaterinvestigation.com/posted/3043/May_27_PDF.670139.pdf.

26 *"If it's a successful test"* "USCG/BOEM Marine Board of Investigation into the Marine Casualty, Explosion, Fire, Pollution, and Sinking of Mobile Offshore Drilling Unit Deepwater Horizon, with Loss of Life in the Gulf of Mexico," April 21–22, 2010, testimony of John Smith, transcript p. 267, lines 1–3, July 23, 2010; http://www.deepwaterinvestigation.com/go/doc/3043/856507/.

26 *"And that was really the only discussion"* "USCG/BOEM Marine Board of Investigation into the Marine Casualty, Explosion, Fire, Pollution, and Sinking of Mobile Offshore Drilling Unit Deepwater Horizon, with Loss of Life in the Gulf of Mexico," April 21–22, 2010, testimony of John Guide, transcript p. 156, lines 13–14, July 22, 2010; http://www.deepwaterinvestigation.com/go/doc/3043/856503/.

27 *"I haven't been a witness"* "USCG/BOEM Marine Board of Investigation into the Marine Casualty, Explosion, Fire, Pollution, and Sinking of Mobile Offshore Drilling Unit Deepwater Horizon, with Loss of Life in the Gulf of Mexico," April 21–22, 2010, testimony of Leo Lindner, transcript pp. 286 and 288, July 19, 2010; http://www.deepwaterinvestigation.com/go/doc/3043/856483/.

27 *" a warning sign right off the bat"* "USCG/BOEM Marine Board of Investigation into the Marine Casualty, Explosion, Fire, Pollution, and Sinking of Mobile Offshore Drilling Unit Deepwater Horizon, with Loss of Life in the Gulf of Mexico," April 21–22, 2010, testimony of John Smith, transcript p. 279, line 10, July 23, 2010; http://www.deepwaterinvestigation.com/go/doc/3043/856507/.

27 *"They attempted it again and got fluid back"* "USCG/BOEM Marine Board of Investigation into the Marine Casualty, Explosion, Fire, Pollution, and Sinking of Mobile Offshore Drilling Unit Deepwater Horizon, with Loss of Life in the Gulf of Mexico," April 21–22, 2010, testimony of Leo Lindner, transcript p. 274, lines 3–4, July 19, 2010; http://www.deepwaterinvestigation.com/go/doc/3043/856483.

27–28 *My supervisor was explaining," and "Where that U-tube's at," and "Bob tells Jason, No," and "approximately a good hour"* "USCG/MMS Marine Board of Investigation into the Marine Casualty, Explosion, Fire, Pollution, and Sinking of Mobile Offshore Drilling Unit Deepwater Horizon, with Loss of Life in the Gulf of Mexico," April 21–22, 2010, testimony of Chris Pleasant, transcript pp. 115–116 and 264, May 28, 2010; http://www.deepwaterinvestigation.com/go/doc/3043/670171.

28 *Don Vidrine believes* J. C. McKinley Jr., "Documents Fill in Gaps in Narrative on Oil Rig Blast," *New York Times*, September 7, 2010, p. A18; http://www.nytimes.com/2010/09/08/us/08rig.html?_r=2&hpw=&pagewanted=all.

28 *"And if you don't see that, you need to be very concerned, right?" and "So the symptoms are a successful test"* "USCG/BOEM Marine Board of Investigation into the Marine Casualty, Explosion, Fire, Pollution, and Sinking of Mobile Offshore Drilling Unit Deepwater Horizon, with Loss of Life in the Gulf of Mexico," April 21–22, 2010, testimony of John Smith, transcript p. 392 and pp. 289–90, July 23,2010; http://www.deepwaterinvestigation.com/go/doc/3043/856507/. *See also* "Video: Deepwater Horizon Joint Investigation Footage, July 23rd, Part 35," posted on July 25, 2010; http://www.dvidshub.net. *See also* "USCG/BOEM Marine Board of Investigation into the Marine Casualty, Explosion, Fire, Pollution, and Sinking of Mobile Offshore Drilling Unit Deepwater Horizon, with Loss of Life in the Gulf of Mexico," April 21–22, 2010, testimony of Lee Lambert, transcript pp. 290, 357, and 395, July 20, 2010; http://www.deepwaterinvestigation.com/go/doc/3043/856499/.

29 *"The investigative team could find no evidence"* BP's Deepwater Horizon Accident Investigation Report, p. 40, September 8, 2010; http://www.bp.com/liveassets/bp_internet/globalbp/globalbp_uk_english/incident_response/STAGING/local_assets/downloads_pdfs/Deepwater_Horizon_Accident_Investigation_Report.pdf.

29 *"Go call the office"* "USCG/MMS Marine Board of Investigation into the Marine Casualty, Explosion, Fire, Pollution, and Sinking of Mobile Offshore Drilling Unit Deepwater Horizon, with Loss of Life in the Gulf of Mexico," April 21–22, 2010, testimony of Chris Pleasant, transcript p. 118, May 29, 2010; http://www.deepwaterinvestigation.com/go/doc/3043/670191/.

30 *Cement bond log test's importance and decision not to use it* D. Hammer, "Costly, Time-Consuming Test of Cement Linings in Deepwater Horizon Rig Was Omitted, Spokesman Says," NOLA.com, May 19, 2010; http://www.nola.com/news/gulf-oil-spill/index.ssf/2010/05/costly_time-consuming_test_of.html.

30 *"Everyone on the rig was completely satisfied"* "USCG/BOEM Marine Board of Investigation into the Marine Casualty, Explosion, Fire, Pollution, and Sinking of Mobile Offshore Drilling Unit Deepwater Horizon, with Loss of Life in the Gulf of Mexico," April 21–22, 2010, testimony of John Guide, transcript p. 43, lines 24–25, July 22, 2010; http://www.deepwaterinvestigation.com/go/doc/3043/856503/.

32 *BP's discussion of Halliburton's cement* BP's Deepwater Horizon Accident Investigation Report, September 8, 2010; http://www.bp.com/liveassets/bp_internet/globalbp/globalbp_uk_english/incident_response/STAGING/local_assets/downloads_pdfs/Deepwater_Horizon_Accident_Investigation_Report.pdf.

32–33 *Halliburton and BP had prior knowledge that the cement failed tests* J. M. Broder, "Panel Says Firms Knew of Cement Flaws Before Spill," *New York Times*, October 28, 2010, p. A1. *See also* D. Cappiello, "Critical Test Not Done on Cement Before Blowout," Associated Press, October 29, 2010. *See also* D. Hammer, "Oil Spill Commission Finds Halliburton's Cement Was Unstable, Failed Several Tests Before Deepwater Horizon Disaster," *Times-Picayune*, October 29, 2010. *See also* "BP, Halliburton Knew Oil Disaster Cement Was Unstable," Agence France-Presse, October 29, 2010.

33 *Even small cracks* PumpCalcs, Orifice flow calculator: http://www.pumpcalcs.com/calculators/view/103/.

33–34 *Events after 8:00 P.M. on April 20* BP's Deepwater Horizon Accident Investigation Report, p. 92, September 8, 2010; http://www.bp.com/liveassets/bp_internet/globalbp/globalbp_uk_english/incident_response/STAGING/local_assets/downloads_pdfs/Deepwater_Horizon_Accident_Investigation_Report.pdf.

34 *"eliminating all conventional well control"* "USCG/BOEM Marine Board of Investigation into the Marine Casualty, Explosion, Fire, Pollution, and Sinking of Mobile Offshore Drilling Unit Deepwater Horizon, with Loss of Life in the Gulf of Mexico," April 21–22, 2010, testimony of John Smith, transcript p. 409, July 23, 2010; http://www.deepwaterinvestigation.com/go/doc/3043/856507/. *See also* "Video: Deepwater Horizon Joint Investigation Footage, July 23rd, Part 49," posted on July 25, 2010; http://www.dvidshub.net.

36 *"Rig crew was not sufficiently prepared"* BP's Deepwater Horizon Accident Investigation Report, p. 108, September 8, 2010; http://www.bp.com/liveassets/bp_internet/globalbp/globalbp_uk_english/incident_response/STAGING/local_assets/downloads_pdfs/Deepwater_Horizon_Accident_Investigation_Report.pdf.

37 *Randy Ezell* "USCG/MMS Marine Board of Investigation into the Marine Casualty, Explosion, Fire, Pollution, and Sinking of Mobile Offshore Drilling Unit Deepwater Horizon, with Loss of Life in the Gulf of Mexico," April 21–22, 2010, testimony of Randy Ezell testimony of Miles Ezell transcript, pp. 283–288, May 28, 2010; http://www.deepwaterinvestigation.com/go/doc/3043/670171.

38 *Mud shooting out the top* BP's Deepwater Horizon Accident Investigation Report, p. 126, September 8, 2010; http://www.bp.com/liveassets/bp_internet/globalbp/globalbp_uk_english/incident_response/STAGING/local_assets/downloads_pdfs/Deepwater_Horizon_Accident_Investigation_Report.pdf.

38 *Alarm inhibition* R. Brown, "Oil Rig's Siren Was Kept Silent, Technician Says," *New York Times*, July 23, 2010 p. A1; http://www.nytimes.com/2010/07/24/us/24hearings.html?_r=1&hp.

39 *"If I would have shut down those engines"* R. Arnold, "Testimony: Safety System 'Inhibited' on Doomed Rig," Click2Houston.com, July 23, 2010; http://www.click2houston.com/news/24373880/detail.html.

39 *Mike Williams's description of events* "Blowout: The Deepwater Horizon Disaster," *CBSNews.com,* May 16,2010; http://www.cbsnews.com/stories/2010/05/16/60minutes/main6490197.shtml. *See also* R. Brown, "Oil Rig's Siren Was Kept Silent, Technician Says," *New York Times,* July 23, 2010, p. A1; http://www.nytimes.com/2010/07/24/us/24hearings.html?_r=1&hp. *See also* "Deepwater Horizon Joint Investigation Hearings Testimony Videos for July 23, 2010," August 4, 2010; http://www.deepwaterinvestigation.com/go/doc/3043/834139/.

40 *Interchange between Pleasant and Kuchta* D. Hammer, "Deepwater Horizon's Ill-Fated Oil Well Could Have Been Handled More Carefully, Hearings Reveal," NOLA.com, May 29, 2010; http://www.nola.com/news/gulf-oil-spill/index.ssf/2010/05/hammer-hearings-cg.html.

40 *Disconnect failed* "USCG/MMS Marine Board of Investigation into the Marine Casualty, Explosion, Fire, Pollution, and Sinking of Mobile Offshore Drilling Unit Deepwater Horizon, with Loss of Life in the Gulf of Mexico," April 21–22, 2010, testimony of Jimmy Harrell, transcript p. 58, May 27, 2010; http://www.deepwaterinvestigation.com/go/doc/3043/670139.

43–44 *Summary of the things that went wrong* "Oil Spill: BP Engineers Partly Responsible," ComodityOnline.com, September 10, 2010; http://www.commodityonline.com/news/Oil-Spill-BP-engineers-partly-responsible-31641-3-1.html.

APRIL

45 *"leading the government response"* C. Woodward, "A Containable Accident, Then Suddenly a Crisis," Associated Press, April 30, 2010.

45–46 *April 23 quotes from Obama, the Coast Guard (except Landry), BP, and Transocean, and the misreported amount of oil* "Major Oil Spill as Rig Sinks off US Coast," Agence France-Presse, April 23, 2010; http://www.physorg.com/news191218312.html.

46 *Michael Brown quotes* D. Sweet, "Michael Brown: Obama Let Oil Spill Get Worse on Purpose," *Raw Story,* May 4, 2010; http://www.rawstory.com/rs/2010/05/michael-brown-obama-wanted-oil-spill.

46 *Obama drilling decision* A. Koppelman, "Is There Anyone Who Likes Obama's Drilling Decision?" Salon.com, March 31, 2010; http://www.salon.com/news/politics/war_room/2010/03/31/drilling_reacts. *See also* J. M. Broder, "Obama to Open Offshore Areas to Oil Drilling for First Time," *New York Times,* March 30, 2010; http://www.nytimes.com/2010/03/31/science/earth/31energy.html.

47–48 *Landry quotes about no apparent leak, plus Robert Gibb and sometimes accidents happen* C. Woodward, "A Containable Accident, Then Suddenly a Crisis,"

Associated Press, April 30, 2010; http://abcnews.go.com/Business/wireStory?id=10513415.

48 *Manageable* F. Grimm, "Vast Oil Spill May Alter Debate on Gulf Drilling," *Miami Herald*, April 27, 2010; http://www.miamiherald.com/2010/04/26/1599498/vast-oil-spill-may-alter-debate.html.

48 *Two to four weeks* S. Mufson, "Oil Spill in Gulf of Mexico Creates Dilemmas in Nature and Politics," *Washington Post*, April 27, 2010; http://www.washingtonpost.com/wpdyn/content/article/2010/04/26/AR2010042604308.html.

50–51 *Relief well, 1,800 square miles, dome, "we have to be careful," "far as the eye could see," off-flavors, and Jindal asked for booms* C. Burdeau, "Oil Leak from Sunken Rig off La. Could Foul Coast," Associated Press, April 26, 2010; http://abcnews.go.com/Business/wireStory?id=10480828.

51 *Relief wells could take two months and that the relief well will cost $100 million* H. Mohr and Cain Burdeau, "Coast Guard Considers Lighting Oil Spill on Fire," Associated Press, April 28, 2010; http://minnesota.publicradio.org/display/web/2010/04/28/oil-spill-gulf/.

51 *Misreported: 4.2 million barrels could spill* C. Burdeau, "La. Gov Declares Emergency over Gulf Oil Spill," Associated Press, April 29, 2010; http://www.salon.com/news/2010/04/29/us_louisiana_oil_rig_explosion_5.

52 *A BP executive agreed on 5,000 barrels a day* "Shrimpers, Fishermen File Suit over US Oil Spill," Agence France-Presse, April 30, 2010; http://www.mysinchew.com/node/38400.

52 *Going from "manageable" to "emergency," Landry's "ample time"* C. Woodward, "A Containable Accident, Then Suddenly a Crisis," Associated Press, April 30, 2010; http://abcnews.go.com/Business/wireStory?id=10513415.

53 *"Shrimpers sue"* "Shrimpers, Fishermen File Suit over US Oil Spill," Agence France-Presse, April 30, 2010; http://www.mysinchew.com/node/38400.

53 *Administration might revisit president's drilling announcement* C. Woodward, "A Containable Accident, Then Suddenly a Crisis," Associated Press, April 30, 2010; http://abcnews.go.com/Business/wireStory?id=10513415.

55 *"Controlled burns," and the relief well will cost $100 million* H. Mohr and Cain Burdeau, "Coast Guard Considers Lighting Oil Spill on Fire," Associated Press, April 28, 2010; http://minnesota.publicradio.org/display/web/2010/04/28/oil-spill-gulf/.

56 *"Eventually, things will return to normal"* H. Mohr and Cain Burdeau, "Gulf Businesses Wait as Oil Creeps Toward Coast," Associated Press, April 29, 2010; http://www.chron.com/disp/story.mpl/business/deepwaterhorizon/6978731.html.

56 *Ixtoc killed hundreds of million of crabs* J. Achenbach and David Brown, "In Gulf Oil Spill's Long Reach, Ecological Damage Could Last Decades," *Washington Post,* June 6, 2010; http://www.washingtonpost.com/wpdyn/content/article/2010/06/05/AR2010060503987.html?hpid=topnews.

DÉJÀ VU, TO NAME BUT A FEW

57 *Histories of antecedent oil spills, in part, from:* Joanna Burger, *Oil Spills,* New Brunswick, NJ: Rutgers University Press, 1997.

58 *Amoco Cadiz* J. Achenbach and David Brown, "In Gulf Oil Spill's Long Reach, Ecological Damage Could Last Decades," *Washington Post,* June 6, 2010; http://www.washingtonpost.com/wp-dyn/content/article/2010/06/05/AR2010060503987.html?hpid=topnews.

59 *"First the money dried up"* A. Allen, "Prodigal Sun," *Mother Jones,* March–April 2000; http://motherjones.com/politics/2000/03/prodigal-sun.

59–60 *Exxon Valdez otter, bird, and seal deaths* C. H. Peterson et al., "Long-Term Ecosystem Response to the *Exxon Valdez* Oil Spill," *Science* 302, December 19, 2003: 2082–86.

60 *Great Britain's Health and Safety Executive warning to Transocean* R. Mason, "Transocean Denies Bullying Oil Rig Workers over Safety Fears," *Daily Telegraph* (London), September 8, 2010; http://www.telegraph.co.uk/finance/newsbysector/energy/7987815/Transocean-denies-staff-bullied-over-safety-fears-in-North-Sea.html.

61 *September 14, 2007, letter* L. Kaufman, "2010 Agency Is Faulted for Saying Oil-Spill Risk to Wildlife Was Low," *International Herald Tribune,* July 6, 2010; http://www.highbeam.com/doc/1P1-181710108.html.

61 *October 2009 environmental assessment* K. Sheppard, "License to Drill," *Mother Jones,* June 2010; http://motherjones.com/environment/2010/06/new-drilling-leases-gulf-of-mexico.

62 *Waxman and Stupak letter to BP* B. Bohrer, "Congressmen Cite 'Significant' BP Issues in Alaska," Associated Press, May 6, 2010; http://abcnews.go.com/Business/wireStory?id=10567540.

62 *Obama's coast plan and the Mississippi Delta* "Obama Administration Officials Release Roadmap for Gulf Coast Ecosystem Restoration Focused on Resiliency and Sustainability," White House Press Release, March 4, 2010; http://www.whitehouse.gov/administration/eop/ceq/Press_Releases/March_4_2010.

62–63 *The senators' letter on energy efficiency and conservation* R. Schoof, "10 Senate Democrats Oppose Climate Bill If It Expands Coastal Drilling," McClatchy-Tribune News Service. March 25, 2010; http://www.mcclatchydc.com/2010/03/25/91107/10-senate-dems-oppose-climate.html.

63 *Chinese-owned coal carrier Shen Neng 1* J. M. Glionna, Ju-min Park, and Kenneth R. Weiss, "Oil Leak Threatens Barrier Reef; Australia Rushes to Protect the Sensitive Ecosystem from a Stranded Chinese Ship Carrying Heavy Fuel," *Los Angeles Times*, April 5, 2010.

63 *Minerals Management Service's Alaska office hasn't developed any guidelines* K. Murphy, "Agency Faulted over Alaska Oil Drilling Policy," *Los Angeles Times*, April 7, 2010; http://articles.latimes.com/2010/apr/07/nation/la-na-alaska-drilling8-2010apr08.

63 *Minerals Management Service approves permit changes* J. Barry, "Oil Spill Disaster Marked by BP Mistakes from the Outset," *St. Petersburg Times*, June 2, 2010; http://www.tampabay.com/news/environment/water/oil-spill-disaster-marked-by-bp-mistakes-from-the-outset/1099273.

64 *"I've never seen this kind of attitude"* "Blowout: The Deepwater Horizon Disaster," CBSNews.com, May 16, 2010; http://www.cbsnews.com/stories/2010/05/16/60minutes/main6490197.shtml.

MAYDAY

67 *"I am frightened for the country," and Crist's "bigger than we can fathom"* C. Woodward, "A Containable Accident, Then Suddenly a Crisis," Associated Press, April 30, 2010; http://abcnews.go.com/Business/wireStory?id=10513415.

68 *"This is the worst possible thing"* H. Mohr and Cain Burdeau, "Coast Guard Considers Lighting Oil Spill on Fire," Associated Press, April 28, 2010; http://minnesota.publicradio.org/display/web/2010/04/28/oil-spill-gulf.

69 *"I'd hate to think" and related quotes* "Oil Leak Poses Risk of a Lost Generation on Louisiana Coast," Agence France-Presse, May 3, 2010; http://www.terradaily.com/reports/Oil_leak_poses_risk_of_a_lost_generation_on_Louisiana_coast_999.html.

69 *Captain Doogie* H. Allen G. Breed and Holbrook Mohr, "Another Week at Least of Unabated Gulf Oil Geyser," Associated Press, May 3, 2010; http://abcnews.go.com/Business/wireStory?id=10531316

69 *"When you kill that food chain," and "We should be fine"* K. McGill, "A Murky Picture as Seafood Industry Eyes Oil Slick," Associated Press, May 3, 2010; http://abcnews.go.com/Business/wireStory?id=10536472.

69 *Volume and value of Gulf catch* "Mobile Baykeeper Says BP Spill Could Destroy Most Productive Fishery in the World," *Energy Weekly News*, May 14, 2010; http://www.prnewswire.com/news-releases/mobile-baykeeper-says-bp-spill-could-destroy-most-productive-fishery-in-the-world-92497804.html.

69–70 *"What BP's doing is throwing," and "It's probably easier to fly," and "It's not a spill, it's a flow," and twenty dead turtles* H. Mohr and Allen G. Breed, "Another

Week at Least of Unabated Gulf Oil Geyser," Associated Press, May 3, 2010; http://abcnews.go.com/Business/wireStory?id=10531316.

71 The Economist *argues, and statistics attributed to the Congressional Research Service and National Research Council* "Deep Trouble: The Oil Spill in the Gulf of Mexico," *The Economist*, May 8, 2010; http://www.economist.com/node/16059948.

71 *Thomas Friedman writes* T. Friedman, "No Fooling Mother Nature," *New York Times.* May 4, 2010; http://www.nytimes.com/2010/05/05/opinion/05friedman.html.

72 *John Kerry quote and Republican governors withdrawing support* C. Babington, "Bid to Enact Energy Bill Might Survive Gulf Spill," Associated Press, May 6, 2010; http://abcnews.go.com/Business/wireStory?id=10569505.

72–73 *Republican climate science denial, and Bob Inglis quoted* "GOP Victory May Be Defeat for Climate Change Policy," NPR's *All Things Considered*, October 23, 2010; http://www.npr.org/templates/transcript/transcript.php?storyId=130776747.

79 *History of tanker wrecks affecting birds* Joanna Burger, *Oil Spills*, New Brunswick, NJ: Rutgers University Press, 1997.

80 *Exxon's CEO didn't care if the booms contained the oil* *Countdown with Keith Olbermann*, MSNBC, June 14, 2010.

80 *"We may . . . lose dozens of vulnerable fish species"* "Dozens of Fish Species in Danger in US Oil Leak," Agence France-Presse, May 31, 2010; http://www.seeddaily.com/reports/Dozens_of_fish_species_in_danger_in_US_oil_leak_expert_999.html.

80 *"We'll see dead bodies soon"* E. Dugan, "Oil Spill Creates Huge Undersea 'Dead Zones,' " *Independent*, May 30, 2010; http://www.independent.co.uk/news/world/americas/oil-spill-creates-huge-undersea-dead-zones-1987039.html.

80 *Could affect rice and sugarcane* R. School and Chris Adams, "Gulf's Future Looks Grim," *Sun News*, May 31, 2010; http://www.thesunnews.com/2010/05/31/1505058/gulfs-future-looks-grim.html.

80 *Hurricane effects* J. Masters, "What Would a Hurricane Do to the Deepwater Horizon Oil Spill?" *Weather Underground*, May 26, 2010; http://www.wunderground.com/blog/JeffMasters/comment.html?entrynum=1492.

80 *The* Christian Science Monitor *asks* P. Jonsson, "Gulf Oil Spill: Could 'Toxic Storm' Make Beach Towns Uninhabitable?," *Christian Science Monitor*, June 26, 2010; http://www.csmonitor.com/USA/2010/0626/Gulf-oil-spill-Could-toxic-storm-make-beach-towns-uninhabitable.

81 *"Once it's in the Loop Current"* C. Burdeau and Harry R. Weber, "Deep Beneath the Gulf, Oil May Be Wreaking Havoc," Associated Press, May 6, 2010; http://abcnews.go.com/Business/wireStory?id=10558496.

81 *Senator Bill Nelson quoted* "Disaster of a New Dimension Looms in Gulf of Mexico Spill," Agence France-Presse, May 9, 2010; http://www.grist.org/article/2010-05-09-disaster-of-a-new-dimension-looms-in-gulf-of-mexico-spill.

81 *"The threat to the deep-sea habitat"* C. Burdeau and Harry R. Weber, "Deep Beneath the Gulf, Oil May Be Wreaking Havoc," Associated Press, May 6, 2010; http://abcnews.go.com/Business/wireStory?id=10558496.

82 *The Web begins amplifying* T. Aym, "How BP Gulf Disaster May Have Triggered a 'World-Killing' Event," Helium.com; http://www.helium.com/items/1882339-doomsday-how-bp-gulf-disaster-may-have-triggered-a-world-killing-event. *See also* T. Aym, "Why BP Is Readying a 'Super Weapon' to Avert Escalating Gulf Nightmare," *Helium.com*; http://www.helium.com/items/1889648-bp-preparing-super-weapon-to-avert-escalating-gulf-nightmare.

83 *Fears not materializing, things not so bad* J. M. Broder and Tom Zeller Jr., "Bad. But an Apocalypse?" *New York Times*, May 4, 2010; http://query.nytimes.com/gst/fullpage.html?res=9A05E7DF113FF937A35756C0A9669D8B63.

83 *Leak could reach 60,000 barrels* J. M. Broder et al., "Amount of Spill Could Escalate, Company Admits," *New York Times*, May 5, 2010; http://www.nytimes.com/2010/05/05/us/05spill.html.

85 *Louisiana's coastal wetlands have lost 2,300 square miles* C. Burdeau and Harry R. Weber, "Deep Beneath the Gulf, Oil May Be Wreaking Havoc," Associated Press, May 6, 2010; http://abcnews.go.com/Business/wireStory?id=10558496.

87 *"It's the last thing in the world"* B. Farrington, "Gulf Spill Gives Gov. Crist Pause over Drilling," Associated Press, April 28, 2010; http://abcnews.go.com/US/wireStory?id=10488587.

87 *"The president is frustrated"* H. J. Hebert and Harry R. Weber, "Political Patience Wanes as Gulf Oil Spill Grows," Associated Press, May 12, 2010; http://www.cbsnews.com/stories/2010/05/12/national/main6475125.shtml.

87 *In 2009 the Interior Department exempted BP's Gulf of Mexico drilling* J. Eilperin, "U.S. Exempted BP Rigs from Impact Analysis; Prior Reviews Concluded That a Large Oil Spill in the Area Was Unlikely," *Washington Post*, May 5, 2010; http://www.washingtonpost.com/wp-dyn/content/article/2010/05/04/AR2010050404118.html.

87 *"Boosterism breeds complacency"* D. Usborne, "US Oil Regulator 'Gave in to BP' over Rig Safety," *Independent*, May 7, 2010; http://www.independent.co.uk/news/world/americas/us-oil-regulator-gave-in-to-bp-over-rig-safety-1965611.html.

88 *Salazar's "very, very rare event"* "Gulf Oil Spill: Interior Secretary Says Spill 'Very, Very Rare Event,' " *Los Angeles Times*, May 2, 2010; http://bit.ly/buSoSo.

88 *39 rig blowouts in the Gulf of Mexico* R. Gold and Ben Casselman, "Drilling Process Attracts Scrutiny in Rig Explosion," *Wall Street Journal,* April 30, 2010; http://online.wsj.com/article/SB10001424052748703572504575214593564769072.html.

88–89 *Sarah Palin* N. Allen, "Foreign Oil Firms Not to Be Trusted, Says Sarah Palin," *Daily Telegraph,* May 7, 2010; http://www.telegraph.co.uk/news/worldnews/northamerica/usa/7686978/Gulf-of-Mexico-oil-slick-Sarah-Palin-fuels-anti-British-sentiment.html.

89 *"I think I have said all along"* D. Usborne, "US Oil Regulator 'Gave in to BP' over Rig Safety," *Independent,* May 7, 2010; http://www.independent.co.uk/news/world/americas/us-oil-regulator-gave-in-to-bp-over-rig-safety-1965611.html.

89 *Transocean's president, Halliburton's spokesman* H. J. Hebert and Harry R. Weber, "Political Patience Wanes as Gulf Oil Spill Grows," Associated Press, May 12, 2010; http://www.cbsnews.com/stories/2010/05/12/national/main6475125.shtml.

89 *Blame among BP, Transocean, and Halliburton* CNN Wire Staff, "Congressman to Launch Inquiry on How Much Oil Is Gushing into Gulf," CNN.com, May 14, 2010; http://articles.cnn.com/2010-05-14/us/gulf.oil.spill_1_bp-oil-spill-dispersants?_s=PM:US.

89–90 *"People from around the world"* "Hair, Fur, Nylons Join Fight to Hold Back US Oil Spill," Agence France-Presse, May 9, 2010; http://www.terradaily.com/reports/Hair_fur_nylons_join_fight_to_hold_back_US_oil_spill_999.html.

90 *Activities in and around Grand Isle* H. R. Weber and John Curran, "Oil Spill Swells to 4M Gallons," Associated Press, May 11, 2010; http://www.cbsnews.com/stories/2010/05/11/national/main6471488.shtml.

90 *"The technology hasn't changed that much"* J. C. McKinley and Leslie Kaufman, "As Oil Spreads, a Call for Industry's 'Plan B,' " *New York Times,* May 11, 2010; http://www.boston.com/news/science/articles/2010/05/11/as_oil_spreads_a_call_for_industrys_plan_b.

90 *"This is the largest, most comprehensive"* H. R. Weber and John Curran, "Oil Spill Swells to 4M Gallons," Associated Press, May 11, 2010; http://www.cbsnews.com/stories/2010/05/11/national/main6471488.shtml.

90 *Workers diverting freshwater* "More River Water Sent into Marshes to Fight Oil," Associated Press, May 10, 2010; http://www.nola.com/news/gulf-oil-spill/index.ssf/2010/05/more_fresh_river_water_sent_in.html.

91 *The looming hurricane season* J. C. Olivera, "Hurricane Could Worsen Huge US Oil Spill," Agence France-Presse, May 12, 2010; http://www.google.com/hostednews/afp/article/ALeqM5jf8KgVCVcikaoZArKPqnF1Rrbheg.

91 *BP's "Safety first"* C. Krauss, "BP Tries Again to Divert Oil Leak with Dome," *New York Times*, May 31, 2010; http://www.nytimes.com/2010/06/01/us/01spill.html.

91–92 *Purdue professor's computer analysis* "Congressman to Launch Inquiry on How Much Oil Is Gushing into Gulf," CNN.com, May 14, 2010; http://articles.cnn.com/2010-05-14/us/gulf.oil.spill_1_bp-oil-spill-dispersants?_s=PM:US.

92 *"If you don't know the flow"* S. Goldenberg, "Scientists Study Ocean Footage to Gauge Full Scale of Oil Leak: Independent Researchers Press BP to Telease Film," *Guardian*, May 14, 2010; http://www.guardian.co.uk/business/2010/may/13/bp-oil-spill-ocean-footage.

LATE MAY

93 *"There's a shocking amount"* J. Gillis, "Giant Plumes of Oil Forming Under Gulf," *New York Times*, May 15, 2010; http://www.nytimes.com/2010/05/16/us/16oil.html.

93 *She reports methane, and "the most bizarre-looking"* M. Brown and Ramit Plushnick-Masti, "Gulf Oil Full of Methane, Adding New Concerns," Associated Press, June 18, 2010; http://www.cbsnews.com/stories/2010/06/18/national/main6594500.shtml.

93 *Deepwater corals* J. Dearen and Matt Sedensky, "Deep Sea Oil Plumes, Dispersants Endanger Reefs," Associated Press, May 17, 2010; http://abcnews.go.com/Business/wireStory?id=10665749.

93–94 *Plumes and related discussion, "The answer is no to that," and "It appears that the subsea dispersant is actually working"* J. Gillis, "Giant Plumes of Oil Forming Under Gulf," *New York Times*, May 15, 2010; http://www.nytimes.com/2010/05/16/us/16oil.html.

94 *Plume science compared to Ixtoc oil* P. D. Boehm and Davld L. Flest, "Subsurface Distributions of Petroleum from an Offshore Well Blowout; The Ixtoc I Blowout, Bay of Campeche," *Environmental Science and Technology* 16, no, 2 (1982): 67–74.

95 *"The answer is no to that"* Justin Gillis, "Giant Plumes of Oil Forming Under the Gulf," *New York Times*, May 15, 2010; http://www.nytimes.com/2010/05/16/us/16oil.html.

96 *Penalties under the Clean Water Act* "The Oil Well and the Damage Done," *The Economist*, June 19, 2010; http://www.economist.com/node/16381032.

96 *"It appears that the application of the subsea dispersant"* Justin Gillis, "Giant Plumes of Oil Forming Under the Gulf," *New York Times*, May 15, 2010; http://www.nytimes.com/2010/05/16/us/16oil.html.

96 *Dispersant making oil available to fish* M. Brown and Jason Dearen, "New, Giant Sea Oil Plume Seen in Gulf," Associated Press, May 27, 2010; http://www.msnbc.msn.com/id/37384717/ns/gulf_oil_spill.

96–97 *Quotes by Shaw, Joye, Overton, and Twilley* "Debate Grows Over Impact of Dispersed Oil," *Nature* magazine, July 10, 2010; http://www.nature.com/news/2010/100710/full/news.2010.347.html.

97 *The mix is more toxic* M. J. Hemmer et al., "Comparative Toxicity of Louisiana Sweet Crude Oil (LSC) and Chemically Dispersed LSC to Two Gulf of Mexico Aquatic Test Species," U.S. Environmental Protection Agency Office of Research and Development, July 31, 2010; http://www.epa.gov/bpspill/reports/phase2dispersant-toxtest.pdf.

98 *Bluefin tuna* "USM Scientists Head Out for Bluefin Tuna Larvae," Associated Press, May 17, 2010; http://www.bettermsreport.com/2010/05/ap-scientists-from-usm-seek-tuna-larvae.

99 *Native Americans* C. Burdeau, "Spill Reinforces Oil Bad Will for American Indians," Associated Press, May 18, 2010; http://abcnews.go.com/US/wireStory?id=10673076.

100 *"We never said it produced 5,000"* M. Winter, "Oil Update: Spill to Harm Europe, Arctic Wildlife; Hurricane Season Looms," *USA Today,* May 21, 2010; http://content.usatoday.com/communities/ondeadline/post/2010/05/oil-update-spill-to-harm-europe-arctic-wildlife-hurricane-season-looms/1.

100 *"We cannot trust BP," and "I don't see any possibility," and "it might be a little more," and "an absurd position"* "BP Accused of Cover-up as Size of Slick Remains Unknown," Agence France-Presse, May 20, 2010; http://www.terradaily.com/reports/BP_accused_of_cover-up_as_size_of_slick_remains_unknown_999.html.

101 *"They want to hide the body"* J. Gillis, "Scientists Fault Lack of Studies over Gulf Spill," *New York Times,* May 20, 2010; http://www.nytimes.com/2010/05/20/science/earth/20noaa.html.

101 *Interview with Admiral Thad Allen* C. Crowley, interview with Thad Allen, *CNN's State of the Union with Candy Crowley,* May 23, 2010; http://transcripts.cnn.com/TRANSCRIPTS/1005/23/sotu.01.html.

101 *Brown and orange globs* K. McGill and Vicki Smith, "A Month in, Outrage over Gulf Oil Spill Grows," Associated Press, May 21, 2010; http://blog.al.com/live/2010/05/a_month_in_outrage_over_gulf_0.html.

101 *Billy Nungesser, and the Louisiana governor, and his statistics, also EPA's deadline* "Louisiana Marshes Hit by Gulf Oil Slick," Agence France-Presse, May 21, 2010; http://www.channelnewsasia.com/stories/afp_world/view/1058133/1/.html.

102 *"This is not sheen"* J. C. McKinley and Campbell Robertson, "Oil Is Fouling Wetlands, Official Says," *New York Times,* May 20, 2010; http://www.nytimes.com/2010/05/20/us/20spill.html.

102 *Dead zone* C. Burdeau, "Dead Zone in the Gulf One of the Largest Ever," Associated Press, August 2, 2010; http://www.cbsnews.com/stories/2010/08/02/tech/main6736957.shtml.

102 *Nancy Rabalais quoted* C. Burdeau and Seth Borenstein, "Scientists Think Gulf Can Recover," Associated Press, August 6, 2010; http://www.msnbc.msn.com/id/38582049/ns/us_news-environment/

103 *Requirements to submit plans, and a "culture of substance abuse and promiscuity* M. Kunzelman and Richard T. Pienciak, "Rule Change Helped BP on Gulf Project," Associated Press, May 6, 2010; http://abcnews.go.com/Business/wireStory?id=10568369.

103 *Salazar dividing the Minerals Management Service* J. C. McKinley and Campbell Robertson, "Oil Is Fouling Wetlands, Official Says," *New York Times,* May 20, 2010; http://www.nytimes.com/2010/05/20/us/20spill.html. *See also* M. Daly, "Minerals Management Service to Be Divided into 3 Parts," Associated Press, May 19, 2010; http://www.huffingtonpost.com/2010/05/19/minerals-management-servi_0_n_582248.html.

104 *"It took hundreds of years"* J. Dearen, "Month into Gulf Spill, Fishermen See Bleak Future," Associated Press, May 21, 2010; http://www.boston.com/news/nation/articles/2010/05/21/month_into_gulf_spill_fishermen_see_bleak_future.

104 *Meanwhile, President Obama announces* B. Walsh, "Obama to Tighten Fuel Economy Etandards," *Time,* May 19, 2010; http://www.time.com/time/health/article/0,8599,1899534,00.html.

104 *"All systems are go"* C. Robertson, "Despite Leak, Louisiana Is Still Devoted to Oil," *New York Times,* May 22, 2010; http://www.nytimes.com/2010/05/23/us/23drill.html.

104–105 *A house divided and related quotes* I. Urbina, "Despite Moratorium, Drilling Projects Move Ahead," *New York Times,* May 23, 2010; http://www.nytimes.com/2010/05/24/us/24moratorium.html. *See also* P. Eckert, "Fishery Disaster Declared in Three Gulf States; U.S. Officials Vow to Impose Fines on Energy Giant," *Ottawa Citizen,* May 25, 2010; Via Reuters; http://www.reuters.com/article/idUSTRE64N5TT20100524. *See also* "BP and the Damage Done," energyboom.com, May 29, 2010; http://www.energyboom.com/policy/bp-and-damage-done.

105 *The president will say he regrets* T. Raum and Jennifer Loven, "Fixing Oil Spill My Responsibility, Obama Says," Associated Press, May 28, 2010; http://abcnews.go.com/Business/wireStory?id=10766709.

106 *Shaw on Corexit 9527* E. Dugan, "Oil Spill Creates Huge Undersea 'Dead Zones,' " *Independent,* May 30, 2010; http://www.independent.co.uk/news/world/americas/oil-spill-creates-huge-undersea-dead-zones-1987039.html.

106 *BP's close ties to Corexit's manufacturer* "Corexit," SourceWatch, August 7, 2010; http://www.sourcewatch.org/index.php?title=Corexit.

106 *Dr. Shaw writes* S. Shaw, "Swimming Through the Spill," *New York Times*, May 28, 2010; http://www.nytimes.com/2010/05/30/opinion/30shaw.html.

107 *Corexit and illness* "Corexit," SourceWatch, August 7, 2010; http://www.sourcewatch.org/index.php?title=Corexit.

107 *Butoxyethanol in Corexit* K. Zeitvogel, "Dispersants: Lesser Evil Against Oil Spill or Gulf Poison?" Agence France-Presse, June 7, 2010; http://www.physorg.com/news195106476.html.

107 *Sarah Palin says Obama's too slow* G. Bluestein and Matt Brown, "Heat on White House to Do More About Gulf Spill," Associated Press, May 25, 2010; http://www.foxnews.com/politics/2010/05/25/heat-white-house-gulf-spill/

108 *The BP response plan* "Did Anyone Actually Read BP's Oil Spill Response Plan?," Targeted News Service, May 25, 2010; http://www.commondreams.org/newswire/2010/05/25-0

108 *Fishermen saying they're getting sick, EPA air monitoring, and related quotes* N. Santa Cruz and Julie Cart, "Oil Cleanup Workers Cite Illness," *Los Angeles Times*, May 26, 2010; http://www.sott.net/articles/show/209299-Gulf-oil-spill-Cleanup-workers-report-illness.

109 *Doug Suttles's "doing everything we can" and Thad Allen's "no news is good"* B. Nuckols and G. Bluestein, "Gulf Awaits Word on Latest Bid to Plug Oil Leak," Associated Press, May 27, 2010; http://www.delcotimes.com/articles/2010/05/27/business/doc4bfe3b8e0c221649103327.txt.

109 *"I wouldn't say it's failed yet," and "They're actually going to take"* "Disaster of a New Dimension Looms in Gulf of Mexico Spill," Agence France-Presse, May 9, 2010; http://www.grist.org/article/2010-05-09-disaster-of-a-new-dimension-looms-in-gulf-of-mexico-spill.

109 *"This is the first time the industry"* J. C. McKinley Jr. and Leslie Kaufman. "New Ways to Drill, Old Methods for Cleanup," *New York Times*, May 10, 2010; http://www.nytimes.com/2010/05/11/us/11prepare.html?_r=2.

110 *$930M and 26,000 people, Hayward quotes* J. Resnick-Ault, "BP Uses 'Junk Shot,' Calls Oil Spill a 'Catastrophe,'" Bloomberg, May 28, 2010; http://www.bloomberg.com/news/2010-05-28/bp-uses-junk-shot-to-plug-well-that-s-spilled-more-oil-than-exxon-valdez.html.

110 *Hayward's "tiny" in a "very big ocean"* T. Webb, "BP Boss Admits Job on the Line over Gulf Oil Spill," *Guardian*, May 14, 2010; http://www.guardian.co.uk/business/2010/may/13/bp-boss-admits-mistakes-gulf-oil-spill.

110 *Discovery of plume by* Weatherbird II M. Brown and J. Dearen, "New, Giant Sea Oil Plume Seen in Gulf," Associated Press, May 27, 2010; http://www.msnbc.msn.com/id/37384717/ns/gulf_oil_spill/.

111 *"This is when all the animals"* E. Dugan, "Oil Spill Creates Huge Undersea 'Dead Zones,' " *Independent,* May 30, 2010; http://www.independent.co.uk/news/world/americas/oil-spill-creates-huge-undersea-dead-zones-1987039.html.

111 *Oil leaked totals between 18 and 39 million gallons* T. Raum and Jennifer Loven, "Fixing Oil Spill My Responsibility, Obama Says," Associated Press, May 28, 2010; http://abcnews.go.com/Business/wireStory?id=10766709.

111 *Obama suspends exploratory drilling in the Arctic* M. Daly, "Salazar Delays Arctic Ocean Drilling," Associated Press, May 27, 2010; http://abcnews.go.com/Business/wireStory?id=10755614.

112 *"This event was set in motion years ago"* R. Simon, "Gulf Oil Spill: Companies Were 'Rushing to Make Money Faster,' Survivor Says," *Los Angeles Times,* May 27, 2010; http://latimesblogs.latimes.com/greenspace/2010/05/gulf-oil-spill-companies-were-rushing-to-make-money-faster-survivor-says.html.

112 *Workers taken ill and arguments over what their problems are* E. Cohen, "Fisherman Files Restraining Order Against BP," CNN, May 31, 2010; http://articles.cnn.com/2010-05-31/health/oil.spill.order_1_bp-oil-rig-oil-spill?_s=PM:HEALTH.

113 *A fisherman's wife, and BP spokesman comment that air monitoring shows no health threats* E. Cohen, "Fisherman's Wife Breaks the Silence," CNN, June 3, 2010; http://articles.cnn.com/2010-06-03/health/gulf.fishermans.wife_1_shrimping-exxon-valdez-oil-spill-cell-phone?_s=PM:HEALTH.

114 *"It's hard to understand if nausea"* F. Tasker and Laura Figueroa, "How Dangerous Is Oil to Human Health?," *St. Petersburg Times,* June 27, 2010.

114 *There's another reason workers aren't using respirators* Elana Schor, "OSHA Chief to Face Hot Seat over Cleanup-Worker Health," *Environment and Energy Daily,* June 21, 2010.

114 *Who else can't breathe?* "Dauphin Island Sea Lab Scientists Report Drop in Oxygen on Alabama Shelf," June 9, 2010; http://press.disl.org/6_9_10oxygenBP.htm.

114–115 *A federal panel of fifty experts* N. Schwartz, "Panel Recommends Continued Use of Oil Dispersant," Associated Press, June 5, 2010; http://abcnews.go.com/Technology/wireStory?id=10828517.

115 *If we don't get as clean a cut"* "Gulf Oil Closes in on Florida, Containment Efforts Hit Snag," Agence France-Presse, June 3, 2010; http://economictimes.indiatimes.com/news/international-business/Gulf-oil-closes-in-on-Florida-containment-efforts-hit-snag/articleshow/6007598.cms.

115 *"Engineer's nightmare," and a tiny cap on a fire hydrant* G. Bluestein and B. Skoloff, "Oil Closes in on Fla. as BP Tries Risky Cap Move," Associated Press, June 2, 2010; http://abcnews.go.com/Business/wireStory?id=10802853.

115–116 *Congress has certainly played a role, and Congress approved budgets reducing regulatory staff, and a 2004 Coast Guard study warned* D. S. Abraham, "A Disaster Congress Voted For," *New York Times*, July 13, 2010, p. A27.

117 *Other nations charge more* J. Wardell and Robert Barr, "BP Not Spearheading Oil Industry Move Back in Gulf," Associated Press, November 2, 2010.

117 *It was actually the Vice President's* National Energy Policy Development Group, *Reliable, Affordable, and Environmentally Sound Energy for America's Future: National Energy Policy Report.* Washington: GPO, 2001.

118 *Congress approved budgets reducing regulatory staff, and a 2004 Coast Guard study warned* D. S. Abraham, "A Disaster Congress Voted For," *New York Times*, July 13, 2010, p. A27.

EARLY JUNE

119 *Dead dolphin, surreptitious tour, and news crews escorted from public places* M. Lysiak and Helen Kennedy, "The Hidden Death in the Gulf," *Daily News*, June 2, 2010; http://www.nydailynews.com/news/national/2010/06/02/2010-06-02_the_hidden_death_in_the_gulf.html.

119 *BP tries again to cover the leak* C. Krauss, "BP Tries Again to Divert Oil Leak with Dome," *New York Times*, May 31, 2010; http://www.nytimes.com/2010/06/01/us/01spill.html?th&emc=th.

120 *Allen's garden hose analogy* H. Mohr and John Flesher, "Gulf Containment Cap Closely Watched in 2nd Day," Associated Press, June 7, 2010; http://www.cbsnews.com/stories/2010/06/06/national/main6553587.shtml.

120 *"A positive step but not a solution"* G. Bluestein, "Cap Placed atop Gulf Well; Oil Still Spewing," Associated Press, June 4, 2010; http://seattletimes.nwsource.com/html/businesstechnology/2012017723_apusgulfoilspill.html.

120 *"The probability of them hitting it"* M. Brown, "Relief for Gulf Is 2 Months Away with Another Well," Associated Press, June 1, 2010; http://abcnews.go.com/Business/wireStory?id=10786704.

120 *BP shares lose, and Justice Department investigations* B. Winter and Kevin Johnson, "Justice Department to Launch Oil Probe," *USA Today*, June 2, 2010; http://www.usatoday.com/news/nation/2010-06-01-criminal-probe-of-oil-spill_N.htm.

120–121 *"The majority, probably the vast majority," and "I don't see that as being a credible," and "They continue to optimize production"* "Scientists Challenge BP Containment Claims," MSNBC, June 9, 2010; http://www.msnbc.msn.com/id/37573643/ns/disaster_in_the_gulf.

122 *"could take leakage almost down to zero," and "I'm not going to declare victory"* "US Sets Deadline for BP as Mistrust Grows," Agence France-Presse, June 10, 2010; http://dalje.com/en-world/us-sets-deadline-for-bp-as-mistrust-grows/308972.

122 *BP shares hemorrhage an incredible 16 percent* F. Ahrens, "BP Stock Crashes; Oil Giant Trading Below Book Value," *Washington Post*, June 9, 2010; http://voices.washingtonpost.com/economy-watch/2010/06/bp_shares_crash_oil_giant_trad.html.

122 *Estimate of the leak gets doubled* "US Doubles Gulf Oil Flow Estimate," Agence France-Presse, June 11, 2010; http://www.dailystar.com.lb/article.asp?edition_id=10&categ_id=2&article_id=115861#axzz16gupLzQJ.

122 *"Still dealing with the flow estimate"* S. Borenstein and Harry R. Weber, "New Oil Numbers May Do More Harm to Fish, Wildlife," Associated Press, June 11, 2010; http://www.fox8live.com/news/local/story/New-oil-numbers-may-do-more-harm-to-fish-wildlife/56mU8Qat2kCx5d5H_7-hSA.cspx.

122 *BP is the third-largest oil company* S. L. Yall and Julie Werdigier, "U.S. Fury at BP Stirs Backlash Among British," *New York Times*, June 10, 2010; http://dealbook.nytimes.com/2010/06/11/u-s-fury-at-bp-stirs-backlash-among-british/.

123 *"I've had guys saying"* B. Anderson, "Nationally, Chefs Stand by Gulf Seafood, but Customers Are Turning Their Backs," *Times-Picayune*, June 20, 2010; http://www.nola.com/news/t-p/frontpage/index.ssf?/base/news-14/1277020369237720.xml&coll=1.

124 *"Thank God there isn't a loaded gun"* S. Saulny, "Cajuns on Gulf Worry They May Need to Move On Once Again," *New York Times*, July 19, 2010; http://bbedit.sx.atl.publicus.com/apps/pbcs.dll/article?AID=/20100719/NEWS0107/7190374/1159&template=print.

124 *Gulf Shores, Alabama* H. Mohr and John Flesher, "Gulf Containment Cap Closely Watched in 2nd Day," Associated Press, June 7, 2010; http://www.cbsnews.com/stories/2010/06/06/national/main6553587.shtml.

124 *Dozens of oil-drenched pelicans* "Dozens of Heavily Oiled Pelicans off La. Coast," Associated Press, June 3, 2010; http://www.fox10tv.com/dpp/news/gulf_oil_spill/heavily-oiled-pelicans-off-la-coast.

124 *More than 35,000 birds* and *approximately 250,000 died* C. H. Peterson et al., "Long-Term Ecosystem Response to the *Exxon Valdez* Oil Spill," *Science* 302, December 19, 2003: 2082–86.

124 *Birds and mammals rescued* "25 Live Sea Turtles Rescued, at Least 12 Oiled," Associated Press, June 2, 2010; http://www.mysanantonio.com/news/environment/25_live_sea_turtles_rescued_at_least_12_oiled_95493429.html.

124–125 *"This is the worst screwed-up response," and quotes by frustrated bird rescuers* C. Pittman, "Bird Experts Left Waiting in Wings," *St. Petersburg Times*, September 5, 2010; http://www6.lexisnexis.com/publisher/EndUser?Action=UserDisplayFullDocument&orgId=574&topicId=100020423&docId=l:1256619892&start=7.

125 *A blowout in Pennsylvania* M. Levy and Jennifer C. Yates, "Marcellus Gas Well Blows Out in Pennsylvania; Gas, Drilling Fluid Shoot 75 Feet into Air," Associated Press, June 4, 2010; http://www.pressconnects.com/article/20100604/NEWS01/6040353/Marcellus-gas-well-blows-out-in-Pennsylvania-gas-drilling-fluid-shoot-75-feet-into-air.

125 *"This spill is just aggregated over a 200-mile radius"* "US Oil Spill Spread Around 200-mile Radius," Agence France-Presse, June 6, 2010; http://www.heraldsun.com.au/news/breaking-news/us-oil-spill-spread-around-200-mile-radius-official/story-e6frf7jx-1225876240262.

126 *BP tells residents in a church* G. Bluestein, "Cap Placed atop Gulf Well; Oil Still Spewing," *Associated Press*, June 2010; http://seattletimes.nwsource.com/html/businesstechnology/2012017723_apusgulfoilspill.html.

126 *Seventy-one people suffer* "Gulf Oil Spill Sickens More Than 70 People in Louisiana," Agence France-Presse, June 9, 2010; http://www.terradaily.com/reports/Gulf_oil_spill_sickens_more_than_70_people_in_Louisiana_999.html.

126 *Department of Health counsels 749 people, and related quotes* M. Navarro, "Spill Takes Toll on Gulf Workers' Psyches," *New York Times*, June 16, 2010; http://www.nytimes.com/2010/06/17/us/17human.html?th&emc=th.

HIGH JUNE

127 *BP good at stopping the flow of news* J. Peters, "Efforts to Limit the Flow of Spill News," *New York Times*, June 9, 2010; http://www.nytimes.com/2010/06/10/us/10access.html?th&emc=th.

130 *Using 5,800 separate measurements* R. Camilli et al., "Tracking Hydrocarbon Plume Transport and Biodegradation at Deepwater Horizon," *Science* 330, no. 6001, August 19, 2010: 201–4; http://www.sciencemag.org/content/330/6001/201.short.

130 *Researchers finding plumes, and BP responses* J. Gillis, "Plumes of Oil Below Surface Raise New Concerns," *New York Times*, June 8, 2010; http://www.nytimes.com/2010/06/09/us/09spill.html.

130 *"Not a river of Hershey's syrup"* "WHOI Scientists Map and Confirm Origin of Large, Underwater Hydrocarbon Plume in Gulf," Woods Hole Oceanographic Institution News Release, August 19, 2010; http://www.whoi.edu/page.do?pid=7545&tid=282&ct=162&cid=79926.

130 *"doesn't hold a candle"* R. Kerr, "Report Paints New Picture of Gulf Oil," *Science*, August 19, 2010; http://news.sciencemag.org/sciencenow/2010/08/report-paints-new-picture-of-gul.html.

131 *NOAA confirmed the presence of undersea oil* "NOAA Completes Initial Analysis of *Weatherbird II* Water Samples," NOAA, June 8, 2010; http://www.noaanews.noaa.gov/stories2010/20100608_weatherbird.html.

131–132 *NOAA's response to undersea plumes, and oil breaking into zillions of droplets* S. Begley, "What the Spill Will Kill," *Newsweek*, June 6, 2010; http://www.newsweek.com/2010/06/06/what-the-spill-will-kill.html.

132–134 *Hayward's "no evidence," "These are huge volumes," and "The bottom line is that yes"* J. Resnick-Ault, "Toxic Plumes Lurk Throughout Gulf," Bloomberg, June 9, 2010; http://www.businessweek.com/news/2010-06-08/toxic-undersea-oil-plumes-lurk-in-gulf-of-mexico-update2-.html.

137 *Energy statistics according to BP* S. Foley, "World Needs Gulf of Mexico's Oil, Says BP," *Independent*, June 10, 2010; http://www.independent.co.uk/news/business/news/world-needs-gulf-of-mexicos-oil-says-bp-1996097.html.

137 *The cap is now collecting 10,000 barrels* "No End in Sight; the Gulf Oil Spill," *The Economist*, June 12, 2010; http://www.economist.com/node/16322752.

138 *Wereley's estimate of 798,000 to 1.8 million gallons* R. Henry et al., "Gulf Oil Leak May Be Bigger Than BP Says," Associated Press, June 9, 2010; http://www.salon.com/news/feature/2010/06/08/gulf_oil_spill_bigger.

138 *Signs warn* M. Nelson, "Florida Posting 1st Signs Warning Swimmers About Oil," Associated Press, June 9, 2010; http://www.foxnews.com/story/0,2933,594242,00.html.

138 *"It's the breadth and complexity"* J. Berger et al., "Dispersal of Oil Means Cleanup to Take Years, Official Says," *New York Times*, June 8, 2010; http://www.nytimes.com/2010/06/08/us/08spill.html.

141 *The futility and utility of helping wildlife* J. Flesher and Noaki Schwartz, "Rescuing Oiled Birds: Poignant, but Is It Futile?," Associated Press, June 10, 2010; http://www.msnbc.msn.com/id/37627833/ns/disaster_in_the_gulf/.

142 *Coated with oil—in Utah* "Oil Coats 300 Birds Near Salt Lake After Pipeline Breach," *Los Angeles Times*, June 13, 2010; http://latimesblogs.latimes.com/greenspace/2010/06/oil-coats-300-birds-near-salt-lake-after-pipeline-breach.html.

LATE JUNE

148 *Oil executives break ranks with BP* J. Broder, "Oil Executives Break Ranks in Testimony," *New York Times*, June 15, 2010; http://www.nytimes.com/2010/06/16/business/16oil.html?ref=politics.

149–150 *BP's profits, and also estimates about oil* P. Nicholas et al., "Stakes Rise for Obama and the Gulf," *Los Angeles Times,* June 15, 2010; http://articles.latimes.com/2010/jun/15/nation/la-na-oil-spill-obama-20100615.

149 *Blowout dampening appetite for greenhouse gas caps* J. Broder, "Oil Spill May Spur Action on Energy, Probably Not on Climate," *New York Times,* June 12, 2010; http://www.nytimes.com/2010/06/13/science/earth/13climate.html.

150 *"That's the $100,000 question," and Orange Beach's mayor* J. Lebovich et al., "BP Promises Swifter Attack Against Oil Spill," *Miami Herald,* June 15, 2010; http://www.miamiherald.com/2010/06/15/1680757/bp-promises-swifter-attack-against.html.

150–151 *A year after the* Exxon Valdez, *and, "I still don't know who's in charge"* C. Robertson, "Efforts to Repel Gulf Spill Are Described as Chaotic," *New York Times,* June 15, 2010; http://www.nytimes.com/2010/06/15/science/earth/15cleanup.html.

151 *Senator Bill Nelson, the mayor of Orange Beach* D. Brooks, "Trim the 'Experts,' Trust the Locals," *New York Times,* June 18, 2010; http://www.nytimes.com/2010/06/18/opinion/18brooks.html.

151 *More than 70 human residents* E. Schor, "Health Consequences of Oil, Dispersant Chemicals Take Center Stage," *Environment and Energy Daily,* June 14, 2010; http://www.eenews.net/eed/2010/06/14.

155 *In mid-June, anger mounts* "Anger Mounts over Moratorium as Obama Tours Spill Zone," Agence France-Presse, June 15, 2010; http://www.lankabusinessonline.lk/fullstory.php?nid=1038762547. *See also* K. Zeitvogel, "Lift 'Reckless' Oil Drilling Ban, Gulf Residents Plead," Agence France-Presse, July 27, 2010; http://www.google.com/hostednews/afp/article/ALeqM5g33v9-JoE_SLNstMjQ5AqOy9DGJw.

156 *Rigs will seek to drill in other countries* T. Zeller Jr., "Drill Ban Means Hard Times for Rig Workers," *New York Times,* June 17, 2010; http://www.nytimes.com/2010/06/18/business/18rig.html.

157 *Also in mid-June, President Obama signals* E. Werner, "Obama: Gulf to Be 'Normal,'" *Associated Press,* June 15, 2010; http://www.huffingtonpost.com/2010/06/14/obama-gulf-coast-visit-th_n_611788.html.

157 *"This was some of the best fishing"* C. Burdeau and Brian Skoloff, "Oil Spill Spreads in La.'s Big, Rich Barataria Bay," Associated Press, June 14, 2010; http://abcnews.go.com/Business/wireStory?id=10909228.

162 *Based on the latest estimates* J. Schwartz, "With Criminal Charges, Costs to BP Could Soar," *New York Times,* June 16, 2010; http://www.nytimes.com/2010/06/17/us/17liability.html?th&emc=th.

163 *The* Economist *reports* "The Oil Well and the Damage Done," *The Economist,* June 19, 2010; http://www.economist.com/node/16381032.

163 *Congressional Democrats Waxman and Markey* "Oil Estimate Raised to 35,000-60,000 Barrels a Day," CNN, June 15, 2010; http://articles.cnn.com/2010-06-15/us/oil.spill.disaster_1_bp-oil-disaster-oil-flow?_s=PM:US.

163 *It will turn out that five giant oil companies, and Further, the companies consistently downplay* S. Mufson and Juliet Eilperin, "Lawmakers Attack Companies' Spill Plans," *Washington Post*, June 16, 2010; http://www.washingtonpost.com/wp-dyn/content/article/2010/06/15/AR2010061501700.html.

164 *Obama's first and much-anticipated address from the Oval Office* "Obama Warns of Oil Spill 'Epidemic' in Oval Office Address," Agence France-Presse, June 16, 2010; http://www.channelnewsasia.com/stories/afp_world/view/1063563/1/.html.

165 *Friedman on Obama and the oil spill* T. Friedman, "Obama and the Oil Spill," *New York Times*, May 18, 2010; http://www.nytimes.com/2010/05/19/opinion/19friedman.html.

165–166 *Rachel Maddow's fake president's speech* "Rachel Maddow Gives Fake Obama Oil Speech, Lunches with Obama," *The Rachel Maddow Show*, MSNBC, June 17, 2010; http://voices.washingtonpost.com/44/2010/06/rachel-maddow-gives-fake-obama.html.

166 *Mary L. Kendall* J. Calmes and Helene Cooper, "BP Chief to Express Contrition in Remarks to Panel," *New York Times*, June 16, 2010; http://www.nytimes.com/2010/06/17/us/politics/17obama.html.

167 *Blair Witherington* K. Murphy, "Death by Fire in the Gulf," *Los Angeles Times*, June 17, 2010; http://articles.latimes.com/2010/jun/17/nation/la-na-oil-spill-burnbox-20100617.

167 *Workers lighting fires at sea* B. Drogin, "Teams Resume Burning Oil in Gulf of Mexico," *Los Angeles Times*, July 12, 2010; http://articles.latimes.com/2010/jul/12/nation/la-na-0712-burn-box-20100712.

168 *People report seeing sharks, mullet* J. Reeves et al., "Sea Creatures Flee Oil Spill, Gather Near Shore," Associated Press, June 17, 2010; http://www.physorg.com/news195978510.html.

168 *BP agrees to pay $20 billion* J. Calmes and Helene Cooper, "BP Chief to Express Contrition in Remarks to Panel," *New York Times*, June 16, 2010; http://www.nytimes.com/2010/06/17/us/politics/17obama.html.

169 *"mind-bogglingly vapid"* "A Bad Day for BP and Mr. Barton," editorial, *New York Times*, June 17, 2010; http://www.nytimes.com/2010/06/18/opinion/18fri3.html.

170 *Oil leaked during the hearing* D. Barry, "Looking for Answers, Finding One," *New York Times*, June 17, 2010; http://www.nytimes.com/2010/06/18/us/18land.html?th&emc=th.

170 *BP releases $25 million* J. Cillis, "New Estimates Raise Amount of Oil Flow," *New York Times*, June 15, 2010; http://www.nytimes.com/2010/06/16/us/16spill.html.

171 *BP's frustration that its acts of goodwill have not been met with goodwill* C. Krauss and John M. Broder, "BP Says Limits on Drilling Imperil Spill Payouts," *New York Times*, September 2, 2010; http://www.nytimes.com/2010/09/03/business/03bp.html.

172 *The total count of sea turtles* "Reporting Oiled Sea Turtles," press release from Sea Turtle Restoration Network, June 18, 2010; http://www.gctts.org/node/557.

173 *"All of these guys could use"* "US Lashes Out at BP as Gulf Oil Spill Reaches Two-Month Mark," Agence France-Presse, June 21, 2010; http://www.hurriyetdailynews.com/n.php?n=us-lashes-out-at-bp-as-oil-spill-reaches-two-month-mark-2010-06-21.

173 *Things Gulf people are saying* K. Murphy, "Oil Spill Stress Begins Taking Its Toll," *Los Angeles Times*, June 20, 2010; http://articles.latimes.com/2010/jun/20/nation/la-na-oil-spill-mental-health-20100621.

174 *Thad Allen had told ABC News* "Allen Has Ordered 'Uninhibited Access' to Oil Spill Operations," ABC News, June 6, 2010; http://bit.ly/904Giz.

174 *The directive in action* M. McClelland, "La. Police Doing BP's Dirty Work," *Mother Jones*, June 22, 2010; http://bit.ly/cFER9Q.

176 *Information on Nigeria* J. Vidal, "Nigeria's Agony Dwarfs the Gulf Oil Spill," *Observer*, May 30, 2010; http://www.guardian.co.uk/world/2010/may/30/oil-spills-nigeria-niger-delta-shell.

177 *Nigerian fraction of U.S. oil imports* "Petroleum," U.S. Energy Information Administration; http://www.eia.gov/oil_gas/petroleum/info_glance/petroleum.html.

177–178 *The* Economist's *estimates, and John Roberts quote* "The Oil Well and the Damage Done," *The Economist*, June 19, 2010; http://www.economist.com/node/16381032.

178 Exxon Valdez's *devastating effects* A. Symington, "Spill Waters Run Deep," *New Statesman*, October 4, 2010; http://www.newstatesman.com/environment/2010/10/oil-spill-liability-gulf.

178 *Obama's May 6 moratorium* M. Kunzelman, "Judge Lifts Offshore Drilling Ban as 'Overbearing,' " Associated Press, June 22, 2010; http://www.heraldsun.com/view/full_story/8016622/article-Judge-lifts-offshore-drilling-ban-as-%E2%80%98overbearing%E2%80%99?instance=homesixthleft.

179 *Perdido and related quotes* J. Mouawad and Barry Meier, "Risk-Taking Rises as Oil Rigs in Gulf Drill Deeper," *New York Times*, August 29, 2010; http://www.nytimes.com/2010/08/30/business/energy-environment/30deep.html.

180 *"There is the pre-April 20th framework of regulation and the post-April 20th framework"* "What Have They Learned?" editorial, *New York Times*, October 5, 2010; http://www.nytimes.com/2010/10/05/opinion/05tue1.html.

181 *The better bet* R. Huval, "Miami Distributor Turns to Imported Shrimp to Replace Gulf's," *Miami Herald*, June 23, 2010; http://www.miamiherald.com/2010/06/23/1695008/miami-distributor-turns-to-imported.html.

181 *Statistics about oysters* "Oil Catastrophe Cracks Gulf of Mexico Oyster Industry," Associated Press, June 23, 2010.

181 *Hurricane Alex strongest since 1966* T. Breen and Jay Reeves, "World's Largest Oil Skimmer Heads to Gulf Spill," Associated Press, July 2, 2010; http://www.foxnews.com/us/2010/07/02/worlds-largest-oil-skimmer-heads-gulf-spill/.

186 *" dumbfounded by the amount of wasted effort"* M. Hennessy-Fiske and Richard Fausset, "Mississippi Officials Blast BP, U.S. Government as Oil Hits Coast," *Los Angeles Times*, June 29, 2010; http://articles.latimes.com/2010/jun/29/nation/la-na-oil-spill-20100629.

186 *"Seeing everything that you've been used to," and "I haven't slept"* J. McConnaughey and Mitch Stacy, "Gulf Oil Spill's Psychological Toll Quietly Mounts as Tragedy Drags On," *Ledger* (Lakeland, FL), June 27, 2010; http://www.theledger.com/article/20100627/news/6275086.

186 *Turtle eggs moved* B. Skoloff, "Some 70,000 Turtle Eggs to Be Whisked Far from Oil," Associated Press, June 30,2010; http://www.physorg.com/news197118448.html.

189 *Where oily absorbent materials are going* F. Barringer, "As Mess Is Sent to Landfills, Officials Worry About Safety," *New York Times*, June 14, 2010; http://www.nytimes.com/2010/06/15/science/earth/15waste.html.

189 *Some of the workforce sits idle* T. Breen, "Volunteers Ready but Left Out of Oil Spill Cleanup," Associated Press, July 2, 2010; http://www.njherald.com/story/news/a1237-BC-US-GulfOilSpill-9thLd-Writethru-07-01-1624.

190 *"The clean-up effort has not been perfect"* "BP Boss Takes Spill Questions in Open Internet Event," Agence France-Presse, July 1, 2010; http://www.muzi.com/news/ll/english/10101675.shtml?q=&cc=26416&a=on.

190 *Appearance of whale sharks* B. Raines, "Whale Sharks Unable to Avoid Oil Spill," Alabama Local News, June 30, 2010; http://blog.al.com/live/2010/06/whale_sharks_unable_to_avoid_o.html. *See also* J. McConnaughey, "Threatened Whale Sharks Seen in Gulf Oil Spill," Associated Press, July 1, 2010; http://www.ajc.com/news/nation-world/threatened-whale-sharks-seen-562061.html.

190 *"the worst possible time" for whale sharks* B. Handwerk, "Whale Sharks Killed, Displaced by Gulf Oil? The Gulf Oil Spill Occurred in Crucial Habitat for the World's Largest Fish," *National Geographic News*, September 24, 2010; http://

news.nationalgeographic.com/news/2010/09/100924-whale-sharks-gulf-oil-spill-science-environment/. *See also* "Whale Shark Tagging in the Gulf of Mexico," a scene from *Mission Blue.* Insurgent Media, June 22, 2010; http://www.youtube.com/watch?v=zVd7aRFaDqY.

190 *Scientists with the University of Southern Mississippi and Tulane University* G. Pender, "Oil Found in Gulf Crabs Raises New Food Chain Fears," *Biloxi Sun Herald*, July 1,2010; http://www.mcclatchydc.com/2010/07/01/96909/oil-found-in-gulf-crabs-raising.html.

190–191 *Effects of oil on fish eggs and larvae* Joanna Burger, *Oil Spills.* New Brunswick, NJ: Rutgers University Press, 1997.

191 *Herring egg and larval mortality following Exxon Valdez* M. D. McGurk and Evelyn D. Brown, "Egg–Larval Mortality of Pacific Herring in Prince William Sound, Alaska, After the Exxon Valdez Oil Spill," *Canadian Journal of Fisheries and Aquatic Science* 53, no. 10 (1996): 2343–54. *See also* B. L. Norcross et al., "Distribution, Abundance, Morphological Condition, and Cytogenetic Abnormalities of Larval Herring in Prince William Sound, Alaska, Following the *Exxon Valdez* Oil Spill," *Canadian Journal of Fisheries and Aquatic Science* 53, no. 10 (1996): 2376 –87. *See also* J. E. Hose et al., "Sublethal Effects of the *Exxon Valdez* Oil Spill on Herring Embryos and Larvae: Morphological, Cytogenetic, and Histopathological Assessments, 1989–1991," *Canadian Journal of Fisheries and Aquatic Science* 53, no. 10 (1996): 2355–65.

191 *Different things get hurt at different rates* C. H. Peterson et al., "Long-Term Ecosystem Response to the *Exxon Valdez* Oil Spill," *Science* 302, no. 5653, December 19, 2003: 2082–86.

192 *Harlequin Ducks still ingesting Exxon Valdez oil* S. Dhillon, "Exxon Oil Showing Up in Alaskan Wildlife 20 years After Spill, Research Shows," *Canadian Press*, April 14, 2010.

192 *Felony announced for getting near booms* C. Kirkham, "Media, Boaters Could Face Criminal Penalties by Entering Oil Cleanup 'Safety Zone,'" *Times-Picayune*, July 1, 2010; http://www.nola.com/news/gulf-oil-spill/index.ssf/2010/07/media_boaters_could_face_crimi.html.

193 *"Never in my lifetime"* G. Nienaber, "Facing the Future as a Media Felon on the Gulf Coast," Huffington Post, July 3, 2010; http://www.huffingtonpost.com/georgianne-nienaber/facing-the-future-as-a-me_b_634661.html.

LIKE A THOUSAND JULYS

196 *BP's dispersant use and cutback* E. Lavandera, "Dispersants Flow into Gulf in 'Science Experiment,'" CNN, July 2, 2010; http://articles.cnn.com/2010-07-02/us/gulf.oil.dispersants_1_corexit-dispersant-bp?_s=PM:US.

196 *EPA and BP's tussle over dispersant policy* E. Rosenthal, "In Standoff with Environmental Officials, BP Stays with an Oil Spill Dispersant," *New York Times*, May 25, 2010; http://www.nytimes.com/2010/05/25/science/earth/25disperse.html.

196 *Corexit 9527 "no longer in use in the Gulf"* E. Schor, "EPA Vows to Push Dispersant Makers After Posting Corexit Ingredients," *Greenwire*, June 10, 2010.

198 *Jon Stewart* "BP's Latest Plan Succeeding, but May Make Spill Worse (for Now)," *Newsweek*, June 2, 2010; http://www.newsweek.com/2010/06/02/first-stage-of-new-bp-plan-succeeds-will-mean-more-oil-in-the-short-term.html#.

198 *BP decriminalizes free speech* J. Reeves, "BP to 40,000 Oil Spill Workers: Talk Away to Media," Associated Press, July 2, 2010; http://abcnews.go.com/US/wireStory?id=11075912.

198 *Relief well very high-tech* H. Fountain, "Hitting a Tiny Bull's-Eye Miles Under the Gulf," *New York Times*, July 5, 2010; http://www.nytimes.com/2010/07/06/science/06drill.html.

199 *The new cap, relief well schedule, Pensacola tourism, and related quotes* F. Tasker, "Officials Offer Glimmer of Hope in Containing Gulf Oil Spill," *Miami Herald*, July 2, 2010; http://www.miamiherald.com/2010/07/02/1712046/rough-seas-still-stymie-oil-spill.html.

200 *Hotline operator calls it a diversion* *Countdown with Keith Olbermann*, MSNBC, June 14, 2010.

202 *Modern corporations are "soulless"* J. Jowit, "'Big Business Needs Biodiversity' Says UN," *Guardian*, July 12, 2010; http://www.guardian.co.uk/environment/2010/jul/12/soulless-corporations-hurt-environment-pavan-sukhdev.

203 *"Arrogant and in denial," and the associated BP management and accident history and related quotes* S. Lyall, "In BP's Record, a History of Boldness and Costly Blunders," *New York Times*, July 12, 2010.

215 *On July 7, the Thadmiral tells America* "Crews Connecting Oil Vessel to Ruptured Well as Leaders Pray for Gulf," AC360 blog, July 7, 2010; http://ac360.blogs.cnn.com/2010/07/07/crews-connecting-oil-vessel-to-ruptured-well-as-leaders-pray-for-gulf/.

216 *Two of the judges on the panel, and appeals court affirms lower court* J. M. Broder, "Court Rejects Moratorium on Drilling in the Gulf," *New York Times*, July 8, 2010; http://www.nytimes.com/2010/07/09/us/09drill.html.

216 *"I'm not going to say you can't drill, but"* M. Daly, "New Drilling Chief Promises Balance," Associated Press, July 12, 2010; http://abcnews.go.com/Business/wireStory?id=11141138.

216 *Salazar issues a new moratorium, and quote from American Petroleum Institute* F. J. Frommer, "Obama Hopes New Drilling Moratorium Can Survive," Associated Press, July 13, 2010; http://www.philly.com/philly/business/20100713_ap_govthopesnewdrillingmoratoriumcansurvive.html?c=0.6657310272741187&posted=n.

217 *the head of the American Petroleum Institute says* "Obama Team Issues New Deepwater Drilling Pause," Frank James, NPR, July 12, 2010; http://npr/igbyaD.

217 *White House senior adviser this week calls the blowout the "greatest environmental catastrophe . . ."* Richard Fausset, "BP Says It's Closer to Oil Containment," *Los Angeles Times*, July 12, 2010; http://lat.ms/gcj9Te.

217 *"greatest environmental catastrophe"* R. Fausset and Nicole Santa Cruz, "Gulf Oil Spill: Containment Cap May Be in Place by Day's End," *Los Angeles Times*, July 12, 2010; http://latimesblogs.latimes.com/greenspace/2010/07/gulf-oil-spill-containment-may-come-by-days-end.html.

220 *News of leaks and BP wanting to leave the well capped* C. Long and Harry R. Weber, "BP, Feds Clash over Reopening Capped Gulf Oil Well," Associated Press, July 19, 2010; http://abcnews.go.com/Business/wireStory?id=11190544.

LATE JULY

222 *Dolphin with broken ribs and a tar ball* S. Dewan, "Sifting a Range of Suspects as Gulf Wildlife Dies," *New York Times*, July 15, 2010; http://www.nytimes.com/2010/07/15/science/earth/15necropsy.html.

222 *Disappearance of birds* C. Burdeau, "Reported Number of Bird Deaths Grow on Gulf Coast," Associated Press, July 23, 2010; http://abcnews.go.com/US/wireStory?id=11231013.

223 *"Just because they kill the well"* H. R. Weber and Greg Bluestein, "BP Begins Effort to Permanently Seal Gulf Gusher," Associated Press, August 4, 2010; http://www.boston.com/news/nation/articles/2010/08/04/bp_begins_effort_to_permanently_seal_gulf_gusher.

223 *"we'll be left to deal with it alone"* G. Bluestein and Tamara Lush, "Crush of Mud Finally Plugs BP's Well in the Gulf," Associated Press, August 5, 2010; http://abcnews.go.com/Business/wireStory?id=11320455.

223 *"We're not going anywhere"* L. Neergaard, " 'CSI' for Seafood: Gulf Fish Gets Safety Tests," *Ledger* (Lakeland, FL), August 16, 2010; http://abcnews.go.com/Health/wireStory?id=11407762.

223 *"They gave us zero"* C. Kirkham, "Seafood Facing a Tainted Image," *Times-Picayune*, August 9, 2010; http://nola.live.advance.net/news/t-p/frontpage/index.ssf?/base/news-15/1281334817130300.xml&coll=1.

224 *An estimated 45,000 people and 6,000 vessels* D. Helvarg, "Obama Signs on for the Sea: A Promising New Commitment to Ocean Protection," *E Magazine*, June 2010; http://www.emagazine.com/view/?5262.

224 *"This whole area is gonna die," and suicide calls* "Faced with Oil Spill, Gulf Residents Fight Mental Pain," Agence France-Presse, July 22, 2010; http://www.rawstory.com/rs/2010/07/faced-oil-spill-gulf-residents-fight-mental-pain.

224 *Chinese pipeline explosion* S. McDonell, "Oil Spews from Exploded China Pipeline," Australian Broadcasting Corporation, July 19, 2010; http://www.abc.net.au/news/stories/2010/07/19/2957868.htm.

224–225 *A federal judge stops companies, and Caroline Cannon quote* D. Joling, "Judge Halts Oil, Gas Development on Chukchi Sea," Associated Press, July 22, 2010; http://www.msnbc.msn.com/id/38352548/.

225 *Daisy Sharp and Steve Oomittuk quoted* "Gulf Spill Focuses Fear of Drilling Disaster in Arctic Villages," *Anchorage Daily News*, August 31, 2010; http://www.adn.com/2010/08/31/1432664/gulf-spill-focuses-fear-of-drilling.html.

225–226 *Oil sands* B. Weber, "Toxic Oil Sands Threat to Fish," *Kamloops Daily News* (British Columbia), August 31, 2010.

226 *Definitions of proppant and fracking fluids* *Schlumberger Oilfield Glossary;* http://www.glossary.oilfield.slb.com.

227 *The New York State Environmental Impact Statement* "Draft Supplemental Generic Environmental Impact Statement on the Oil, Gas and Solution Mining Regulatory Program," *New York State Department of Environmental Conservation*, October 26, 2009; http://www.dec.ny.gov/energy/58440.html.

227 *The final EPA Pavillion report* "Pavillion Area Groundwater Investigation, Pavillion, Fremont County, Wyoming," United States Environmental Protection Agency: Contract No. EP-W-05-050, TDD No. 0901-01, see p. 37, August 30, 2010; http://www.epa.gov/region8/superfund/wy/pavillion/Pavillion_GWInvestigationFSP.pdf.

228 *"It's doubtful we will ever use it"* J. Mouawad, "4 Oil Firms Commit $1 Billion for Gulf Rapid-Response Plan," *New York Times*, July 21, 2010; http://www.nytimes.com/2010/07/22/business/energy-environment/22response.html.

228 *Sixteen months before completion of containment equipment* H. R. Weber, "Oil Industry Has Yet to Adopt Lessons of BP Spill," Associated Press, October 4, 2010; http://www.timesherald.com/articles/2010/10/05/news/doc4caab57b10006826476498.txt.

228–230 *Gaylord Nelson, Santa Barbara, Earth Day, Lois Capps, and the Associated Press poll on Americans and the environment* F. J. Frommer, "Gulf Spill Lacks Societal Punch of Santa Barbara," Associated Press, July 29, 2010; http://abcnews.go.com/Business/wireStory?id=11275571.

231 *Seafood sniffers* Y. Q. Mui and David A. Fahrenthold, "Gulf Seafood Must Pass the Smell Test; Government Trains Inspectors to Sniff Out Contaminated Catch," *Washington Post*, July 13, 2010; http://www.highbeam.com/doc/1P2-25347383.html.

231 *"Take small bunny sniffs," and "If you think about your ability to detect," and sniff-related quotes* K. Lohr, "Gulf Fish Cordoned Off for Seafood Sniffers' Inspection," NPR's *Weekend Edition*, July 18, 2010; http://www.npr.org/templates/story/story.php?storyId=128600385.

233 *PAHs in seafood and oysters at the top of the concern list* "Gulf Seafood Undergoes Intense Testing," Associated Press, August 16, 2010; http://www.huffingtonpost.com/2010/08/16/gulf-seafood-undergoes-in_n_682935.html.

233–234 *Federal regulators say they're sure, and Dawn Nunez quoted* J. Dearen and Greg Bluestein, "Gulf Seafood Declared Safe; Fishermen Not So Sure," Associated Press, August 2, 2010; http://www.rdmag.com/News/FeedsAP/2010/08/energy-gulf-seafood-declared-safe-fishermen-not-so-sure.

233 *The 3,500-plus samples were safe* D. A. Fahrenthold and Juliet Eilperin, "Unallayed by Tests, Fishermen Greet Start of Gulf Shrimp Harvest with Suspicion," *Washington Post,* August 17, 2010; http://www.washingtonpost.com/wp-dyn/content/article/2010/08/16/AR2010081605471.html.

233–234 *"It tastes like fish," and "These are decent American people"* C. Robertson, "In Gulf, Good News Is Taken with Grain of Salt," *New York Times,* August 4, 2010; http://www.nytimes.com/2010/08/05/us/05gulf.html.

234 *"I wouldn't feed that to my children"* D. A. Fahrenthold and Juliet Eilperin, "Unallayed by Tests, Fishermen Greet Start of Gulf Shrimp Harvest with Suspicion," *Washington Post,* August 17, 2010; http://www.washingtonpost.com/wp-dyn/content/article/2010/08/16/AR2010081605471.html.

234 *"Will there even be a market"* B. Boxall et al., "BP Jams Gulf Well with Drilling Mud," *Los Angeles Times,* August 5, 2010; http://articles.latimes.com/2010/aug/05/nation/la-na-oil-spill-20100805.

234 *"People are scared of it right now"* S. Dewan, "Questions Linger as Shrimp Season Opens in Gulf," *New York Times,* August 17, 2010; http://www.nytimes.com/2010/08/17/us/17gulf.html.

234 *Barge crashes into inactive well* R. Adams, "Barge Hits Abandoned Oil Well off Louisiana Coast," *Wall Street Journal,* July 27, 2010; http://online.wsj.com/article/SB10001424052748703977004575393672718018794.html.

235 *"University scientists have spotted"* G. Pender, "Oil Found in Gulf Crabs Raises New Food Chain Fears," *Biloxi Sun Herald,* July 1, 2010; http://www.mcclatchydc.com/2010/07/01/96909/oil-found-in-gulf-crabs-raising.html.

236 *BP quarterly loss of $16.9 billion, and 3,800 miles of boom* A. Gully, "100 Days in, Gulf Spill Leaves Ugly Questions Unanswered," Agence France-Presse, July 28, 2010; http://www.google.com/hostednews/afp/article/ALeqM5invct2tdMai2ExWeQaGJwFmdt9Xg.

DOG DAYS

239 *Origins of oil* W. J. Broad, "Tracing Oil Reserves to Their Tiny Origins," *New York Times,* August 3, 2010; http://www.nytimes.com/2010/08/03/science/03oil.html.

240 *History of early oil uses and extraction* Joanna Burger, *Oil Spills*, New Brunswick, NJ: Rutgers University Press, 1997.

242 *Potential in algae* "Slimy Scum May Hold the Key to Freedom from Oil," Editorial: *Canberra Times*, August 4 2010; http://www.canberratimes.com.au/news/opinion/editorial/general/slimy-scum-may-hold-the-key-to-freedom-from-oil/1903354.aspx.

242 *Benefits of algae as fuel* R. H. Wijffels and M. J. Barbosa, "An Outlook on Microalgal Biofuels," *Science* 329, no. 5993 (2010): 796–99.

246–247 *"I think it is fairly safe to say"* S. Borenstein, "Looking for the Oil? NOAA Says It's Mostly Gone," Associated Press, August 5, 2010; http://abcnews.go.com/Technology/wireStory?id=11328193.

248 *Jim Cowan quoted* B. Boxall et al., "BP Jams Gulf Well with Drilling Mud," *Los Angeles Times*, August 5, 2010; http://articles.latimes.com/2010/aug/05/nation/la-na-oil-spill-20100805.

248–249 *Georgia Sea Grant's alternative report* A. Miles, "Scientists Wary of U.S. Report That Says Only 26 Percent of Spilled Gulf Oil Left," *Times-Picayune*, August 17, 2010; http://www.nola.com/news/gulf-oil-spill/index.ssf/2010/08/scientists_wary_of_us_report_t.html.

249 *Ronald J. Kendall quoted, and Lubchenco's "I think the view"* B. Boxall et al., "BP Jams Gulf Well with Drilling Mud," *Los Angeles Times*, August 5, 2010; http://articles.latimes.com/2010/aug/05/nation/la-na-oil-spill-20100805.

249 *"The truth is in the middle"* D. Fahrenthold and Leslie Tamura, "Majority of Spilled Oil in Gulf of Mexico Unaccounted for in Government Data," *Washington Post*, July 29, 2010; http://www.washingtonpost.com/wp-dyn/content/article/2010/07/28/AR2010072806135.html.

249 *"Roughly one-third of the oil"* A. Miles, "Scientists Wary of U.S. Report That Says Only 26 Percent of Spilled Gulf Oil Left," *Times-Picayune*, August 17, 2010; http://www.nola.com/news/gulf-oil-spill/index.ssf/2010/08/scientists_wary_of_us_report_t.html.

249 *Only about 26 percent of the oil is still in the water or onshore, and Carol Browner quote* J. Gillis, "U.S. Finds Most Oil from Spill Poses Little Additional Risk," *New York Times*, August 4, 2010; http://www.nytimes.com/2010/08/04/science/earth/04oil.html.

249–250 *"Only about one-quarter of the oil," and Lubchenco's "Dilute and out of sight"* S. Connor, "How Did Five Million Barrels of Oil Simply Disappear?" *Independent*, August 6, 2010; http://www.independent.co.uk/news/science/how-did-five-million-barrels-of-oil-simply-disappear-2044817.html.

250 *Bill Lehr statement* S. Goldenberg, "US Scientist Retracts Assurances over State of BP Spill," *Guardian*, August 20, 2010; http://www.guardian.co.uk/environment/2010/aug/19/bp-oil-spill-scientist-retracts-assurances.

250 *Quotes from McDonald, Condrey, and Ault* B. Marshall, "Don't Kid Yourself, Oil Isn't Gone," *Times-Picayune*, August 8, 2010; http://www.nola.com/sports/t-p/index.ssf?/base/sports-47/1281336612181230.xml&coll=1.

251 *"We're going to have to think about what to do"* "BP Says It Might Drill Again in Gulf Reservoir," Associated Press, August 6, 2010; http://www.usatoday.com/communities/greenhouse/post/2010/08/bp-oil-spill-gulf-1/1.

252 *"People died out there," and BP hedging about relief wells* H. R. Weber, "To Seal or Sell? BP Has Options on Remaining Oil," Associated Press, August 6, 2010; http://abcnews.go.com/Business/wireStory?id=11328201.

252 *"I am the national incident commander"* H. R. Weber and Greg Bluestein. "BP Begins Effort to Permanently Seal Gulf Gusher," Associated Press, August 4, 2010; http://www.boston.com/news/science/articles/2010/08/04/bp_begins_effort_to_permanently_seal_gulf_gusher/.

252 *BP hedging about relief wells and Allen's response* J. Dearen and Greg Bluestein, "Crude Still Coats Marshes and Wetlands Along Gulf," Associated Press, August 6, 2010; http://www.signonsandiego.com/news/2010/aug/06/crude-still-coats-marshes-and-wetlands-along-gulf/. *See also* "BP Says It Might Drill Again in Gulf Reservoir," Associated Press, August 6, 2010; http://www.usatoday.com/communities/greenhouse/post/2010/08/bp-oil-spill-gulf-1/1.

253 *Transocean asks for a liability limit of $27 million* "Partner Says BP Hiding Oil Spill Documents," Agence France-Presse, August 19, 2010; http://www.google.com/hostednews/afp/article/ALeqM5hE2vTYfDJqNnNj8r3reXXT6fZ7wg.

254 *Gulf is well populated with bacteria* T. C. Hazen et al. "Deep-Sea Oil Plume Enriches Indigenous Oil-Degrading Bacteria," *Science* 330, no. 6001, August 24, 2010: 204–8.

254 *Ronald Atlas and Roger Sassen quoted* W. J. Broad, "Oil Spill Cleanup Workers Include Many Very, Very Small Ones," *New York Times*, August 4, 2010; http://www.nytimes.com/2010/08/05/science/earth/05microbe.html.

255 *Floating oil mats largely gone, and Lubchenco quotes* J. Gillis and Campbell Robertson, "On the Surface, Oil Spill in Gulf Is Vanishing Fast," *New York Times*, July 27, 2010; http://www.nytimes.com/2010/07/28/us/28spill.html.

257 *Fraud alert* D. Usborne, "BP Believes 10% of Oil Spill Claims Are Fraudulent," *Independent*, August 12, 2010; http://www.independent.co.uk/news/world/americas/bp-believes-10-per-cent-of-oil-spill-claims-are-fraudulent-2050309.html.

257–258 *Bob Shipp and Mark Davis quoted* T. Smith, "By Hiring Gulf Scientists, BP May Be Buying Silence," National Public Radio, July 31, 2010; http://www.npr.org/templates/story/story.php?storyId=128892441&sc=17&f=1001.

LATE AUGUST

259 *BP wondering if bottom kill is necessary* T. Breen and Harry R. Weber, "Gulf Leaders Wary over Wavering on Final Plug," Associated Press, August 12, 2010; http://www.indianexpress.com/news/gulf-leaders-wary-over-wavering-on-final-plu/659395/.

259–260 *"A bottom kill finishes this well," and "a hard sell"* T. Breen, "Decision Expected on Plug for BP's Broken Oil Well," Associated Press, August 13, 2010; http://abcnews.go.com/Business/wireStory?id=11390638.

261 *No oil-caused dead zones developed* J. Raloff, "No 'Dead Zone' from BP Oil," *Science News*, September 7, 2010; http://www.sciencenews.org/view/generic/id/63103/title/No_dead_zone_from_BP_oil.

262 *Sargassum recovering* M. Hirsch, "Fewer Turtles Being Found Soaked in Oil, Researchers Say," *Times-Picayune*, August 11, 2010; http://www.nola.com/news/gulf-oil-spill/index.ssf/2010/08/fewer_turtles_being_found_soak.html. *See also* L. Kaufman and Shaila Dewan, "Luck and Ecology Seem to Soften Impact of Gulf Oil Spill," *International Herald Tribune*, September 14, 2010; http://www.highbeam.com/doc/1P1-184098210.html.

262 *"The Gulf is incredible"* "Nearly 80 Percent of Gulf Spill Oil Still in Water: Experts," Agence France-Presse, August 18, 2010; http://www.independent.co.uk/environment/nearly-80-percent-of-gulf-spill-oil-still-in-water-experts-2055910.html.

262 *Oil and dispersant droplets in the larval crabs* M. Thomas, "Oil-in-Seafood Reports; Do They Hold Water?" *Orlando Sentinel*, August 12, 2010.

262 *"The spill isn't as bad," and "The perception is"* C. Robertson, "In Gulf, Good News Is Taken with Grain of Salt," *New York Times*, August 4, 2010; http://www.nytimes.com/2010/08/05/us/05gulf.html.

263 *Green shoots appearing, and Associated Press calculates 3.4 square miles* C. Burdeau and Jeffrey Collins, "Signs of Regrowth Seen in Oiled Louisiana Marshland," Associated Press, August 12, 2010; http://www.boston.com/news/science/articles/2010/08/12/signs_of_regrowth_seen_in_oiled_louisiana_marshland/.

263 *"We were anchored"* B. Marshall, "Don't Kid Yourself, Oil Isn't Gone," *Times-Picayune*, August 8, 2010; http://www.nola.com/sports/t-p/index.ssf?/base/sports-47/1281336612181230.xml&coll=1.

264 *Very roughly 1,800 square miles* K. Wells, "Collapsing Marsh Dwarfs BP Oil Blowout as Ecological Disaster," *BloombergBusinessweek*, August 18, 2010; http://www.businessweek.com/news/2010-08-18/collapsing-marsh-dwarfs-bp-oil-blowout-as-ecological-disaster.html. *See also* S. Borenstein and Cain Burdeau, "Scientists Think Gulf Can Recover," Associated Press, August 6, 2010; http://abcnews.go.com/Technology/wireStory?id=11337200.

264 *Marsh losses* J. W. Day Jr. et al., "Restoration of the Mississippi Delta: Lessons from Hurricanes Katrina and Rita," *Science* 315, no. 5819, March 23, 2007: 1679–84. *See also* L. D. Britsch and Joseph B. Dunbar, "Land Loss Rates: Louisiana Coastal Plain," *Journal of Coastal Research* 9, no. 2, Spring 1993: 324–38; http://www.jstor.org/stable/4298092.

264 *Felicia Coleman quoted* C. Robertson, "Gulf of Mexico Has Long Been Dumping Site," *New York Times,* July 30, 2010; http://query.nytimes.com/gst/fullpage.html?res=9A04E4D8133AF933A05754C0A9669D8B63.

264 *Mississippi marsh loss* C. Burdeau and Jeffrey Collins, "Signs of Regrowth Seen in Oiled Louisiana Marshland," Associated Press, August 12, 2010; http://www.boston.com/news/science/articles/2010/08/12/signs_of_regrowth_seen_in_oiled_louisiana_marshland/.

265 *Navy secretary Ray Mabus* H. R. Weber and Dina Cappiello, "Obama Endorses Using Fines for Gulf Rehabilitation," Associated Press, September 29, 2010; http://news.yahoo.com/s/ap/20100929/ap_on_bi_ge/us_gulf_oil_spill_restoration.

265 *Louisiana shrimpers go back to work* D. A. Fahrenthold and Juliet Eilperin, "Unallayed by Tests, Fishermen Greet Start of Gulf Shrimp Harvest with Suspicion," *Washington Post,* August 17, 2010; http://www.washingtonpost.com/wp-dyn/content/article/2010/08/16/AR2010081605471.html.

265 *BP's money for mental health* "Plumes of Gulf Oil Spreading East on Sea Floor," AC360 blog, August 17, 2010; http://ac360.blogs.cnn.com/2010/08/17/plumes-of-gulf-oil-spreading-east-on-sea-floor.

266 *Margaret Carruth, a Louisiana husband and wife, and "throwing desks, kicking chairs"* J. Reeves, "Oil Gusher Is Dead, but Not Residents' Anguish," Associated Press, September 27, 2010; http://news.yahoo.com/s/ap/20100927/ap_on_re_us/us_gulf_oil_spill_mental_health.

266 *Allen wanting to avoid timelines* H. R. Weber, "Feds: No Timeline for Completing Gulf Relief Well," Associated Press, August 18, 2010; http://www.physorg.com/news201364019.html.

267 *Beluga whales* A. Mayeda, "Whale Sanctuary Open to Oil Drilling; Small Portion Put Aside for Exploration," *Calgary Herald,* August 28, 2010; http://www.calgaryherald.com/technology/Whale+sanctuary+open+drilling/3454789/story.html.

267 *Indonesia demands $2.4 billion* "Indonesia Demands 2.4 Billion Dollar Payout over Oil Spill," Agence France-Presse, August 31, 2010; http://www.thenewage.co.za/Detail.aspx?news_id=288&cat_id=1026&mid=53.

267 *Thai-owned company rejects Indonesia's demand* "Thai Firm Rejects Indonesia's 2.4-Billion Oil Spill Claim," Agence France-Presse, September 3, 2010; http://www.thejakartaglobe.com/home/thai-firm-rejects-indonesias-24-billion-oil-spill-claim/394363.

EARLY SEPTEMBER

269 *"Yes!" the sign says* R. Fausset, "Shrimp and Oil Are Still King at This Louisiana Festival," *Los Angeles Times,* September 6, 2010; http://articles.latimes.com/2010/sep/06/nation/la-na-petroleum-festival-20100906.

269 *Oil near Panaji, India* "Goa on Slippery Slope as Mystery Ship Dumps Oil," *Pioneer* (India), September 2, 2010; http://www.dailypioneer.com/280374/Goa-on-slippery-slope-as-mystery-ship-dumps-oil.html.

270 *"We are not finding anything"* M. Newsom, "Officials Say No Submerged Oil Found," *Biloxi Sun Herald,* September 10, 2010.

270 *"We haven't seen any oiled sediments"* C. Burdeau, "NOAA Says Sediment on Gulf Floor Not Visibly Oiled," Associated Press, September 30, 2010; http://www.wkrg.com/gulf_oil_spill/article/noaa-says-sediment-on-gulf-floor-not-visibly-oiled/994300/Sep-30-2010_3-30-pm/.

270 *"I've never seen anything like this"* R. Harris, "Scientists Find Thick Layer of Oil on Seafloor," National Public Radio, September 10, 2010; http://www.npr.org/templates/story/story.php?storyId=129782098.

270 *"It wasn't like a drape"* "Experts See Oil in Gulf as Threat to Marine Life," *Los Angeles Times,* August 18, 2010.

270–271 *Jindal's berms* C. Burdeau, "EPA: Louisiana's Sand Berms Not Stopping Much Oil," Associated Press, September 10, 2010; http://www.dailycomet.com/article/20100910/FEATURES12/100919996.

273 *Raising the failed blowout preventer* H. R. Weber, "Engineers Prepare to Remove Gulf Well's Cap," Associated Press, August 28, 2010. *See also* H. R. Weber, "Risks Remain with Gulf Well Cap Coming Off," Associated Press, September 2, 2010; http://news.yahoo.com/s/ap/20100902/ap_on_bi_ge/us_gulf_oil_spill.

273 *John Wright, and drilling the final stretch* J. Collins, "Gulf Relief Well Down to Final, Tricky 100 Feet," Associated Press, August 9, 2010; http://www.scpr.org/news/2010/08/09/gulf-relief-well-down-final-tricky-100-feet.

THE NEW LIGHT OF AUTUMN

274 *The cement stake* A. G. Breed, "The Well Is Dead, but Gulf Challenges Live On," Associated Press, September 19, 2010.

285 *Freshwater largely demolished Louisiana's oysters* J. C. Rudolf, "For Oysters, a 'Remedy' Turned Catastrophe," *New York Times,* July 21, 2010; http://green.blogs.nytimes.com/2010/07/21/for-oysters-a-remedy-turned-catastrophe/.

285 *Dead deep-water corals* J. C. Rudolf, "Dead Coral Found Near Site of Oil Spill," *New York Times,* November 5, 2010; http://www.nytimes.com/2010/11/06/science/earth/06coral.html.

287 *"effectively dead"* "BP Oil Well Declared 'Dead.' " *Washington Post,* September 21, 2010; http://www.washingtonpost.com/wp-dyn/content/article/2010/09/20/AR2010092005835.html.

287 *"like a shadow out there"* C. Morgan, "Study: Oil from Spill Has Not Become a Drifting Cloud of Death," *Miami Herald,* September 8, 2010; http://www.miamiherald.com/2010/09/07/1813032/study-oil-from-spill-has-not-become.html.

287 *"Red snapper are unbelievable"* S. Borenstein and Cain Burdeau, "Scientists Lower Gulf Health Grade," Associated Press, October 19, 2010; http://news.yahoo.com/s/ap/20101019/ap_on_sc/us_gulf_survival.

287 *What are scientists finding?* B. Raines, "Researcher: Fish Numbers Triple After Oil Spill Fishing Closures," *Mobile Register,* November 7, 2010; http://blog.al.com/live/2010/11/researcher_fish_numbers_triple.html.

289 *More than 27,000 abandoned wells* J. Donn and Mitch Weiss, "Gulf Awash in 27,000 Abandoned Wells," Associated Press, July 7, 2010; http://www.msnbc.msn.com/id/38113914/ns/disaster_in_the_gulf.

291 *The oil has everyone's attention* A. Revkin, "While Oil Gushes, Invisible Ocean Impacts Build," Dot Earth blog, *New York Times;* http://dotearth.blogs.nytimes.com/2010/06/18/while-oil-gushes-invisible-ocean-impacts-build/.

292 *The heart and lungs of the planet* D. Smith, "Alarm at Speeding Sea Change," *Age* (Melbourne, Australia), June 18, 2010; http://newsstore.theage.com.au/apps/viewDocument.ac?page=1&sy=age&kw=deborah+smith&pb=all_ffx&dt=selectRange&dr=6months&so=relevance&sf=text&sf=author&sf=headline&rc=10&rm=200&sp=nrm&clsPage=1&docID=AGE100618C649UFM4JOH.

292 *Science has just published a series* J. Smith et al., "Changing Oceans," *Science* 328, no. 5985, June 18, 2010: 1497.

295 *Drawbacks of renewable energy sources, and "they don't offer new services"* "Scaling Up Alternative Energy," special section of *Science* magazine 329, August 13, 2010: 779–803.

296 *A CBS News/New York Times poll* "Poll: Vast Majority Say U.S. Energy Policy Needs Major Changes," CBCnews, June 21, 2010; http://www.cbsnews.com/8301-503544_162-20008368-503544.html.

298–299 *Gannets* "BP Spill Threatens a Third of Canadian Gannets," CBCnews, October 21, 2010; http://www.cbc.ca/technology/story/2010/10/21/nl-montevecchi-gannets-1021.html.

299 *"oil everywhere"* S. Davis, "La. Coast Hit by More Oil," *Advocate,* September 25, 2010.

ACKNOWLEDGMENTS

Mark Loehr's herculean devotion to getting the story of what happened in the well and on the rig absolutely right has immensely benefitted both this book and me. As a researcher, he stands in a class of his own. I believe it is likely that for most of the summer Mark understood the totality of what happened to cause the blowout, the specific details of events, and the role of different personnel in the various companies better than any other single person, period. I could never have truly afforded him.

On the science side, John Angier was exceptionally generous, prompt, and entertainingly attuned to debunking hype and hysteria and setting records straight. I thank also William Semple for his expertise and insight.

Admiral Thad Allen, the national incident commander, and Dr. Jane Lubchenco, the chief of the National Oceanic and Atmospheric Administration—neither of whom had time to spare and both of whom had much better things to do—generously gave me two hours of their time, during which they efficiently conveyed months of experience and years of wisdom.

In and around the Gulf region, I benefited from the generous assistance of the National Wildlife Federation's Karla Raettig, Amanda Moore, Emily Guidry Schatzel, and Larry Schweiger.

Jo Billups, a singer and activist, was extremely generous in helping arrange my aerial perspective and in welcoming me into her home and community. I also thank filmmaker Bill Mills, Marion Laney, Frank Campo, Charlie Robin, Jeff Wolkart, James Fox, Gary Skinner, Reverend Chris Schansberg, and Julian MacQueen of the Hampton Inn in Pensacola Beach. George Brower graciously hosted me on my first night in New Orleans. I thank Campbell Robertson for introducing me to Oliver Houck. For other logistical assistance, I thank the Gulf Restoration Network, especially Jonathan Henderson. And photographer Jeffrey Dubinsky. Jennifer Godwin of the Suncoast Seabird Sanctuary, near Tampa, told me about the gannets.

All season long, my understanding of the Gulf-wide picture benefitted greatly from the constant stream of reporting from the dedicated professional journalists at the *New York Times*, the Associated Press, the *Washington Post*, the *Los Angeles Times*, Agence France-Presse, and elsewhere, whose importance to society is now so undervalued. If not for the news they uncover and deliver, there'd be little to huff about. I thank Connie Murtagh of SeaWeb for sending hundreds of articles my way.

Sylvia Earle and Charlotte Vick provided continual encouragement. I was first persuaded to overcome my initial reluctance to go to the Gulf by Steve Dishart and Myra "Just go" Sarli. As always, the indefatigable and always reliable Megan Smith arranged and backstopped my logistics. I thank also Kate McLaughlin, Alan Duckworth, and the staff of Blue Ocean Institute for their patience in my absence. Jesse Bruschini provided feedback and feed. For interest and encouragement and discussion, I thank the faculty and staff of Stony Brook University's School of Marine and Atmospheric Sciences and School of Journalism's Center for Communicating Science. I thank Judy Bergsma for introducing me to Mark Loehr. When I bemoaned my deadline, Joanna Burger provided her usual "you can do it" encouragement. I thank Jack Macrae for his graciousness, Jean Naggar for her diligence, and John Glusman for his confidence.

For their interest in my impressions and opinions, I thank congressional representatives Ed Markey, Lois Capps, and their able staffs, as well as the hosts and staffs of *Democracy Now*, PBS's *Need to Know*, CNN, and, particularly, Richard Galant, MSNBC's Keith Olbermann, *Audubon* magazine, the Blue Ocean Film Festival, the organizers and participants at the TEDxOilSpill conference, and Stephen Colbert and the producers of *The Colbert Report.*

I thank my family, Patricia and Alexandra, for dealing with a serially difficult summer on the home front. Being seventeen is not for the faint of heart, but as the saying goes, what doesn't kill us makes us stronger, and Alex has made us proud. Things were at times more difficult because of my absences and, at times, because of my presence. Not least, while I was in the Gulf, Pat had to deal alone with a middle-of-the-night decision to end the life of our dog, Kenzie. Sad though that was, we were again touched by the healing powers of the new young furred and feathered creatures who came—and literally fell—into our lives this year. Each time I returned home drained after confronting the grief in the Gulf, those young lives were my smile factory. The topography of life includes much rough terrain. When big things go off the rails, it's the little things that keep one sane.

INDEX

Abandoned oil wells, 289–291
Alaska oil and gas development, halting of, 224–225
Alaska pipeline oil spills, 61, 116
Alexander the Great, 240
Algae as source of oil, 242
Allen, Adm. Thad, x, 101, 125–126, 138, 182
  amount of spilled oil, estimates of, 96, 122, 150, 281
  BP's behavior during oil spill crisis, assessment of, 279–280
  criminalization of boom safety-zone violations, 276–277
  Deepwater Horizon closure after blowout, 104–105, 109–110, 115, 120, 121, 122, 199, 215, 220, 252, 259, 260, 266, 287
  dispersant use in cleanup effort, 277–278, 283
  hero status, 275–276
  hurricane concerns, 181
  media coverage of Gulf oil spill, 174
  Safina's interview with, 275–286
Amoco Cadiz oil spill, 58
Anderson, Jason, 14, 27, 28–29, 42
Anderson, Shelley, 182
Apache Corporation, 290
Arctic oil drilling, 267
Argo Merchant oil spill, 58, 191
Atlas, Ronald M., 254
Ault, Jerald, 250
Baran, Janet, 270
Barbour, Haley, 138, 186, 188
Bartlit, Fred H., Jr., 33
Barton, Joe, 169
Bassey, Nnimo, 177
Bea, Robert, 179–180
Bertone, Steve, 23
Birds, 77–79, 104, 124–125, 141, 222, 298–299
Block, Melissa, 201
Bly, Mark, 23
Boehner, John, 47
Book of Heroic Failures (Pile), 58
BP oil company, 5, 228
  Alaska pipeline oil spill, 61
  Allen's assessment of BP's behavior during oil spill crisis, 279–280
  amount of spilled oil, estimates of, 92, 95, 100–101, 121
  anti-foreign sentiment directed at, 123
  dividend payout issue, 137, 168
  financial status, 71, 120, 122, 177, 186, 221, 236
  Hayward's replacement with Dudley, 172–173
  honorable behavior toward Gulf residents, 267–268
  legal strategy for blowout lawsuits, 253, 257–258
  media coverage of Gulf oil spill and, 127–128, 136, 174–176, 198

BP oil company (*cont.*)
  mental-health program for Gulf residents, 265, 286
  resumed drilling at Deepwater Horizon site, consideration of, 251–252
  safety policy and record, 62, 64, 162–163, 198, 203–205
  status in oil industry, 122–123, 203
  Texas City refinery explosion, 60, 162–163, 203, 204
  See also Deepwater Horizon closure after blowout; Deepwater Horizon oil blowout; Oil spill cleanup effort
Brown, Douglas, 24, 39
Brown, Michael, 46, 276
Browne, John, 203–204
Browner, Carol, 247
Burgess, Michael, 170
Burkeen, Aaron Dale, 42
Bush, George W., 74–75, 117

Campaign financing, 75
Campo, Frank, 183–184
Cannon, Caroline, 225
Capps, Lois, 137, 229, 230
Carden, Stan, 40, 41
Carruth, Margaret, 266
Carter, Jimmy, 59
Center for Biological Diversity, 47
Chasis, Sarah, 230
Cheney, Dick, 117
Chow, Edward C., 179
Chu, Steven, 165
Clark, Donald, 42
Clean fuels, 293–294, 295–296, 298
Climate change, 62–63, 72–75, 150, 203, 229
Coast Guard, 45, 151, 172, 192–193
Cocales, Brett, 19–20
Coleman, Felicia, 264
Colwell, Rita, 170
Congress, 169–170, 243
  climate-change legislation, 62–63, 150, 229
  energy legislation, 71, 72, 164
  oil taxes and royalties, 116–117
  politicizing of Gulf disaster, 149–150
  regulation of oil industry, 118, 170–171
"Conservative" phonies, 188–189
Cooney, Philip, 74–75
Coral reefs, 285
Corexit 9527 dispersant, 106, 107, 196
Corporations
  legal status as persons, 129–130
  multinationals, threat posed by, 297
  responsibilities of, 217–218
  UN report on, 202–203
Costner, Kevin, 153
Cowan, Jim, 248
Crist, Charlie, 67, 70, 72, 87
Crowley, Candy, 101
Crozier, George, 114, 232
Cullen, Nathan, 267
Curtis, Steven Ray, 8, 37, 42
Cuyahoga River fire, 55

Darce, Lee, 269
Davis, Kevin, 197–198
Davis, Mark, 258
Davis, W. Eugene, 216
Dead zone in Gulf of Mexico, 102
Deepwater Horizon closure after blowout
  blowout preventer removal, 273
  BP/government responsibility issue, 104–105
  cap operation of July (static kill), 218–220, 259
  collection of oil in tankers, 120–122, 137, 148–149, 218–219, 220
  dome-pump-pipe-ship system, 50–51, 70–71
  failure of attempted techniques in earlier blowout, 116
  junk shot operation, 108, 110
  permanent plugging of well, 274, 287
  plane ride over ground zero, 154–155
  relief well operation (bottom kill), 51, 70, 120, 198–199, 215, 219, 230, 236, 252–253, 259–260, 266, 273, 274
  remotely operated vehicles and, 281
  response plan lacking at time of blowout, 48, 87–88, 108, 150–151

sealing off of well, 236
stopping of leak, 220
time requirements, 48, 51, 199, 266, 281–282
top hat operation, 115–116, 119–120
top kill operation, 109–110
tube-in-leaking-pipe operation, 99–100
Deepwater Horizon oil blowout
alarm systems and, 38–39
blame-trading by companies, 89, 253
blowout events, 3–4, 37–42
blowout preventer and, 8, 25, 36–37, 38, 271
BP's report on, 43–44, 271–272
casualties, 4, 42–43, 48
cement job and, 8, 16–22, 24, 29–33
centralizers and, 19–21, 31–32
as chain disaster, 44
cost considerations and, 12–13
criminal and civil investigations into, 120
displacing of heavy fluids with seawater and, 29, 33–36
emergency disconnect system and, 40
exploratory well's conversion into production well and, 16, 27
families of killed workers, companies' treatment of, 182
float collar and, 33
history of problems at Deepwater Horizon, 8–9, 10–11
kill-pill material in spacer and, 22–23, 25, 28, 34, 36
liability for, 33
long string casing design and, 11–13, 18
managerial changes and, 13–16
oil executives' views on, 148
permit changes and, 63
pressure tests and, 24–29
research program to study effects of, 170
survivors' treatment by companies, 112
warnings prior to blowout, 14
See also Deepwater Horizon closure after blowout; Oil spill in Gulf of Mexico
Deepwater Horizon oil platform, 4–5, 9
Deepwater oil drilling, 4, 5–8, 61
DeMint, Jim, 243–244
Dispersant use in cleanup effort, 54, 96–97, 102, 105–107, 113, 114–115, 131, 132, 133, 196–198, 245, 277–278, 283
Dmytryk, Rebecca, 124, 125
Dolphins, 54, 119, 124, 157, 167–168, 193, 222, 245, 256, 285
Drake, Edwin, 241, 295
Dudley, Bob, 173, 190, 205

Earle, Sylvia, 276, 288–289
Emoto, Masaru, 171
Energy legislation, 71, 72, 164
Environmental degradation due to fossil fuel use, 291–293
Environmental movement, 228–230
Environmental Protection Agency (EPA), 102, 106–107, 196, 197, 227, 271, 290
Eustice, Brittin, 218
Exxon Valdez oil spill, 52, 55–56, 80, 106, 111, 122, 125, 142, 196, 220, 251, 253
damages penalty against Exxon, 177–178
events of, 59–60
long-term impact, 191–192
seafood industry and, 223
wildlife impact, 60, 124, 191–192
Ezell, Randy, 37–38, 40–41

Federal Aviation Administration (FAA), 127
Ferguson, Patty, 99
Fish and Wildlife Service, 61, 124, 186
Fisher, Charles, 285
Fisheries closures, 53, 99, 111, 123, 177, 181, 199–201, 231, 232, 257, 265, 266, 269, 287–288
Fleytas, Andrea, 41, 42
Florida oil spill, 58
Ford, Henry, 194–195

Fossil fuel use, environmental consequences of, 291–293
Fox, Lee, 124–125
Fracking technique for gas extraction, 226–227
Friedman, Thomas, 71, 165
Fuel-efficiency standards, 104

Gagliano, Jesse, 17, 18, 19, 20
Gas fracking, 226–227
Gasoline prices, 198
Genghis Khan, 240
Georgia Sea Grant program, 248–249
Gibbs, Robert, 47, 107, 247
Gillespie, Tommy, 142, 145–146
Government Accountability Office (GAO), 63, 290
Graham, Lindsey, 62, 150, 229
Graham, Monty, 114
Greenpeace, 47, 104, 270
Groundwater contamination, 227
Guide, John, 10, 12–13, 17, 19, 20, 21, 26, 30

Hafle, Mark, 10–11, 13, 18
Halliburton company, 5, 30, 32–33, 89, 253
Harper, Stephen, 267
Harrell, Jimmy, 14, 17, 23, 24, 40
Hayward, Tony, 89, 90, 95, 101, 109, 110, 113, 116, 120–121, 132, 149, 172–173, 203, 251, 291
  congressional testimony, 169–170
Heck, Ken, 288
Helvarg, David, 157
Hoffmayer, Eric, 190
Holcomb, Jay, 125
Hollander, David, 111, 270
Homeland Security Department, 128
Houck, Oliver, 242–243

Ikari, Ben, 177
Inglis, Bob, 73–74
Interior Department, 87, 103, 118
Ixtoc I oil blowout, 56, 58–59, 83, 94, 116, 196, 219, 253–254

Jackson, Lisa, 105–106, 107, 196
Jindal, Bobby, 51, 101, 105, 152, 172, 215, 270–271, 285
Johnson, Paul, 15
Johnson, Ron, 72
Jones, Gordon, 43
Joye, Samantha, 93, 94, 96, 130, 270
Justice Department, 120

Kalamazoo River oil spill, 235
Kaluza, Bob, 13, 15, 23, 27, 28, 29
Kemp, Roy Wyatt, 42–43
Kendall, Mary L., 166
Kendall, Ronald J., 249
Kennedy, David, 67
Kennedy, Keith, 97, 98
Kerosene, 241
Kerry, John, 62, 72, 164
Kieff, Casey, 76
Kieff, James, 79–80
Kleinow, Kevin, 96
Kleppinger, Karl, Jr., 43
Kuchta, Curt, 14, 40

Ladner, Keath, 234
Lambert, Lee, 29
Landrieu, Mary, 156
Landry, Rear Adm. Mary, 45–46, 48, 52, 53, 55, 280
Laney, Marion, 205–206, 210–211, 213
Lautenberg, Frank, 47
Lehr, Bill, 250
Lieberman, Joe, 62, 164
Lindner, Leo, 15, 23, 27
Loop Current, 81, 82, 83–84, 201
Lubchenco, Jane, x, 131, 133–134, 223, 249–250, 255
  Safina's interview with, 275–286
Lysiak, Matthew, 128

Mabus, Ray, 265
MacDonald, Ian, 100, 133, 250

Macondo formation, 3, 18
Maddow, Rachel, 116, 165–166
Manuel, Keith Blair, 43
Marine sanctuaries, 288–289
Marine Spill Response Corporation, 90
Markey, Edward, 87, 100, 163, 243
Marshall, Bob, 263–264
Marshland destruction, 84–85, 263–265
Marshland impact of Gulf oil spill and cleanup, 53, 90, 101, 263, 279
Matte, Tim, 269
McCain, John, 46, 47
McClelland, Mac, 175
McConnell, Mitch, 173
McCormack, Greg, 13
McCurdy, Cody, 142–143, 144, 145, 146–147
McKinney, Larry, 111
McNutt, Marcia, 111
Media coverage of Gulf oil spill, 71, 119, 127–128, 136, 172, 174–176, 186, 194, 198, 235, 249–250
Miller, Corey, 153, 154, 155
Mills, Bill, 134
Minerals Management Service (MMS), 26, 61, 63, 88, 150, 163, 166, 214, 290
  corrupt practices within, 103
Montevecchi, Bill, 299
Moore, Mandy, 182–183, 186
Moratorium on deepwater drilling, 111, 155–156, 178, 180, 215–217, 228
Morel, Brian, 13, 16, 18, 19
Muir, David, 172
Murawski, Steve, 84, 287
Murray, Chad, 40, 41

National Energy Policy, 117
National Oceanic and Atmospheric Administration (NOAA), 131, 133, 231, 232, 233, 266, 269, 270, 287
  report on fate of oil in Gulf, 246–250
National Oil Spill Contingency Plan, 278
Native Americans, 99, 224–225
Nelson, Bill, 63, 81, 128, 151
Nelson, Gaylord, 228, 229
Nigerian oil spills, 176–177
Nixon, Richard, 229
Norris, Michele, 201
Nuclear energy, 294–295
Nunez, Dawn, 234
Nungesser, Billy, 53, 101, 105, 151, 216

Obama, Barack, 75, 87, 105, 107, 128, 152, 156–157, 229
  BP escrow account to cover spill-related damages, 168–169
  first reactions to Gulf oil spill, 45, 52
  fuel-efficiency policy, 104
  Gulf Coast plan, 62
  Gulf Coast visits, 149
  moratorium on deepwater drilling, 111, 155, 178, 180, 228
  offshore oil development policy, 46–47, 53
  opportunity presented by Gulf oil spill, failure to exploit, 164–166
Occupational Safety and Health Administration (OSHA), 114, 204
Oil industry, 194–195
  disaster response plans, 163–164
  history of oil extraction and use, 239–242
  moratorium on deepwater drilling, 111, 155–156, 178, 180, 215–217, 228
  Obama's offshore oil development policy, 46–47, 53
  oilsands exploitation, 225–226
  political influence, 242–243
  regulation of, 103, 117–118, 166, 170–171, 180, 188–189, 216, 228
  risks of deepwater oil drilling, disregard for, 178–180
  safety shortcomings, 9–10
  standby capability for future blowouts, establishment of, 227–228
  subsidies for, 294
Oil Pollution Act of 1990, 150, 267, 277, 278, 286
Oil spill cleanup effort
  beach cleanups, 135, 138, 189
  BP escrow account to cover spill-related damages, 149, 157, 168–169
  BP's expenditures on, 110, 199, 289

Oil spill cleanup effort (*cont.*)
BP's hotline for suggestions, 199
BP's response plan, 108
BP's responsibility for, 53–54
BP's self-interest in, 143
burning off of oil, 55, 148–149, 167
closing of public places as "secure areas," 207–210
containment booms, 51, 78, 80, 85, 86, 91, 144, 192–193, 277
credibility problem of BP and government agencies, 197, 223, 246, 247, 260
criminalization of boom safety-zone violations, 192–193, 216, 276–277
disaggregation of oil and, 138
disappearance of oil over time, 287
dispersant use, 54, 96–97, 102, 105–107, 113, 114–115, 131, 132, 133, 196–198, 245, 277–278, 283
health problems among workers, 108, 112–115, 224
local residents hired for, 91, 142–144, 194, 210–212
love and appreciation approach, 171–172
marshland impact, 90, 279
microbes' consumption of oil, 223, 254, 283
Mississippi River diversion to protect coast, 90, 285
outsiders brought to Gulf Coast for, 136
public discontent with, 151
sand berm creation, 90, 206, 207, 271
skimming operations, 244–245
skimming vessel from Taiwan for, 187–188
underused workforce, 189–190, 210, 257
Vessels of Opportunity program, 210
wildlife rescues, 124–125, 141, 152, 186–187, 207
Oil spill in Gulf of Mexico
amount of spilled oil, 48, 51–52, 54, 83, 91–92, 93, 95–96, 100–101, 111, 121, 122, 125, 137–138, 150, 196, 219, 220–221, 281
Barataria Bay impact, 157–158, 263
boat ride through affected waters, 144–147
catastrophizing related to, 81–83, 217, 291
first reactions by government, 45–46, 47–48, 52, 280–281
fisheries closures, 53, 99, 111, 123, 177, 181, 199–201, 231, 232, 257, 265, 266, 269, 287–288
Grand Isle impact, 159–162
Gulf residents' reactions, 49–50, 68–69, 79–80, 123–124, 126, 134–135, 139–141, 158, 162, 173, 182–186, 194–195, 210–215
hurricane concerns, 80, 91, 181
less damage than expected resulting from, 261–263, 270, 287–288, 291
lessons learned issue, 272
long-term impact, 55–56, 284
Loop Current concerns, 81, 82, 83–84, 201
marshland impact, 53, 101, 263
media coverage of, 71, 119, 127–128, 136, 172, 174–176, 186, 194, 198, 235, 249–250
Mississippi River mouth impact, 97–98
NOAA report on fate of oil, 246–250
oxygen depletion problem, 114–115, 131, 261–262
plane ride over affected waters, 153–155
plant life impact, 166–167, 262
plumes of undersea oil, 93–95, 110–111, 130–134
politicizing of, 149–150, 229
protests regarding, 139
psychological and social impacts, 251, 260–261, 265–266, 285–286
public health concerns, 134–135, 138, 140, 151
religious perspectives on, 136–137, 205
seafood industry and, 69, 99, 101–102, 123, 181, 211, 223, 224, 231, 235, 261 (*See also* fisheries closures *above*)
siege mentality caused by, 67–68, 82

slicks on water surface, 85–86
tourism industry and, 138, 140, 141, 207, 261
U.S. environmental policy and, 229–230
wildlife impact, 70, 77–79, 80, 81, 98, 104, 111, 114–115, 119, 124, 132, 151, 157, 167–168, 172, 190–192, 193, 214, 218, 222–223, 235, 250–251, 254–257, 262, 283–285, 287–288, 298–299
See also Oil spill cleanup effort
Oil spills of the past, 57–64, 79, 116, 176–177, 191–192, 228, 229. *See also Exxon Valdez* oil spill; Ixtoc I oil blowout
Oil spill threats of the future, 289–291
Oil taxes and royalties, 116–117
Oomittuk, Steve, 225
Osterholm, Michael, 114
Otavie, Bonny, 177
Ott, Riki, 142
Overpopulation problem, 298
Overton, Ed, 96–97, 262
Oysters, 285

Palin, Sarah, 88–89, 107, 123
Panepinto, Rhonda, 127
Perdido oil platform, 179
Perry, Harriet, 190
Petrae, Gary, 270
Pfeiffer, Yvonne, 186
Picou, Steven, 251
Pile, Stephen, 58
Piltz, Rick, 74
Piper Alpha oil blowout, 59
Pleasant, Chris, 27, 28, 39–40
Population explosion, 298
Powers, Sean, 287–288
Prudehoe Bay oil spill, 61, 204

Rabalais, Nancy, 102
Reagan, Ronald, 59, 216
Reddy, Chris, 130
Renewable energy sources, 293–294, 295–296, 298
Rensink, David, 120
Revette, Dewey, 14, 43
Riley, Bob, 172
Roberts, John, 178
Robin, Charlie, 184–185
Robin, Doogie, 69
Roshto, Shane, 43

Salazar, Ken, 88, 103, 104, 180, 216–217
Santa Barbara oil spill, 57, 133, 228, 229
Sargassum weed, 166–167, 262
Sassen, Roger, 254
Sawyer, Diane, 172
Schansberg, Chris, 136–137
Schmalz, Sharon, 125
Schwarzenegger, Arnold, 72
Schweiger, Larry, 217, 218
Scientific ignorance in America, 72–75
Sea Empress oil spill, 60
Seafood industry, 69, 99, 101–102, 123, 181, 211, 223, 224, 231, 235, 261. *See also* Fisheries closures
Sepulvado, Ronnie, 16
Sharks, 190, 257, 287–288
Sharp, Daisy, 225
Shaw, Susan, 96, 106
Shen Neng I oil spill, 63
Shipp, Bob, 257–258
Sims, David, 12, 16, 23
Skinner, Gary, 211–212
Smith, Jerry E., 216
Smith, John R., 26, 27, 28, 34
Solar energy, 59, 295
Stacy, Brian, 262
Stern, Nicholas, 203
Stevens, John Paul, 75
Stewart, Jon, 198
Stone, Stephen, 112
Stupak, Bart, 62, 149, 169, 204
Sukhdev, Pavan, 202–203
Supreme Court, 75, 129–130, 177–178
Suttles, Doug, 109, 251

Taylor, Caz, 235
Tebbit, Lord, 123
Texas City refinery explosion, 60, 162–163, 203, 204

Timor Sea oil blowout, 267
Tjeerdema, Ronald S., 56
Torrey Canyon oil spill, 57, 191
Toups, Dicky (Captain Coon-ass), 84, 85, 86
Tourism industry, 138, 140, 141, 207, 261
Trahan, Buddy, 41
Transocean company, 3, 5, 14, 33, 48, 60, 89, 112, 116, 182, 253
Treibs, Alfred E., 239
Tuna, 98, 255
Turtles, 70, 78, 119, 141, 151, 167, 172, 207, 222, 262, 265, 284–285
Twilley, Robert, 97

United Nations, 202–203

Valentine, John, 287, 288
Verdin, Sydney, 99
Vidrine, Don, 13, 23, 27, 28, 29

Walz, Gregg, 19
Wathen, John, 157, 167–168
Watt, James, 103
Waxman, Henry A., 62, 100–101, 149, 163, 169, 204
Weise, Adam, 43
Weiss, Daniel, 171
Wereley, Steve, 91–92, 100, 121, 138
Wetlands. *See* Marshland *headings*
Whales, 225, 267
Wheelan, Andrew, 174–175
Wheeler, Wyman, 27, 40, 41
Wildlife impact of Gulf oil spill, 70, 77–79, 80, 81, 98, 104, 111, 114–115, 119, 124, 132, 151, 157, 167–168, 172, 190–192, 193, 214, 218, 222–223, 235, 250–251, 254–257, 262, 283–285, 287–288, 298–299
Wildlife rescue efforts, 124–125, 141, 152, 186–187, 207
Wildlife Response Services, 125
Williams, Mike, 24, 38, 39, 41–42, 43
Wind energy, 293, 295
Witherington, Blair, 167, 262
Wolkart, Jeff, 193
Wright, John, 273

Yeats, W. B., 189
Yellow Sea oil spill, 224

ABOUT THE AUTHOR

Carl Safina has studied the ocean as a scientist, stood for it as an advocate, and conveyed his travels among the sea's creatures and fishing people in lyrical nonfiction. His first book, *Song for the Blue Ocean*, was chosen as a *New York Times* Notable Book of the Year, and in 2000 he won the Lannan Literary Award for nonfiction. Dr. Safina's second book, *Eye of the Albatross*, won the John Burroughs Medal for the year's best nature book; it was chosen by the National Academies of Science, Engineering, and Medicine as the "Year's Best Book for Communicating Science." The *New York Times* selected both his *Voyage of the Turtle* as well as *The View from Lazy Point: A Natural Year in an Unnatural World* as "Editors' Choices." Safina is founding president of Blue Ocean Institute and adjunct professor at Stony Brook University, where he is involved with its School of Marine and Atmospheric Sciences and its Center for Communicating Science. He has been named among "100 Notable Conservationists of the Twentieth Century" by *Audubon* magazine.